STUDENT SOLUTION

ESSENTIALS OF GENERAL CHEMISTRY

SECOND EDITION

Ebbing/Gammon/Ragsdale

DAVID BOOKIN
MT. SAN JACINTO COLLEGE

DARRELL D. EBBING
WAYNE STATE UNIVERSITY

STEVEN D. GAMMON
WESTERN WASHINGTON UNIVERSITY

RONALD RAGSDALE
UNIVERSITY OF UTAH

HOUGHTON MIFFLIN COMPANY ■ BOSTON ■ NEW YORK

Vice President and Publisher: Charles Hartford
Executive Editor: Richard Stratton
Development Editor: Kellie Cardone
Editorial Associate: Rosemary Mack
Senior Project Editor: Claudine Bellanton
Senior Art/Design Coordinator: Jill Haber
Manufacturing Coordinator: Carrie Wagner
Senior Marketing Manager: Katherine Greig
Marketing Coordinator: Alexandra Shaw

Printed in the U.S.A.

ISBN: 0-618-49178-3

23456789-MP-09 08 07 06 05

CONTENTS

PREFACE

This student solutions manual provides worked-out answers to the problems that appear in *Essentials of General Chemistry, 2ⁿᵈ Edition*, by Darrell D. Ebbing, Steven D. Gammon, and Ronald Ragsdale. This includes detailed, step-by-step solutions for the odd-numbered Practice Problems, General Problems, and Cumulative-Skills Problems, that appear at the end of the chapters. Also provided are answers to all the Review Questions. No answers are given for the Concept Checks and Conceptual Problems.

Please note the following:

Significant figures: The answer is first shown with 1 to 2 nonsignificant figures and no units, and the least significant digit is underlined. The answer is then rounded off to the correct number of significant figures, and the units are added. No attempt has been made to round off intermediate answers, but the least significant digit has been underlined wherever possible.

Great effort and care have gone into the preparation of this manual. The solutions have been checked and rechecked for accuracy and completeness several times. I would like to express my thanks to Stephen Z. Goldberg of Adelphi University for his assistance in accuracy reviewing.

D. B.

ESSENTIALS OF GENERAL CHEMISTRY

SECOND EDITION

STUDENT SOLUTIONS MANUAL

1. CHEMISTRY AND MEASUREMENT

■ Answers to Review Questions

1.1   One area of technology that chemistry has changed is the characteristics of materials. In devices such as watches and calculators, the liquid-crystal displays (LCDs) are materials made of molecules designed by chemists. Electronics and communications have been transformed by the development of optical fibers to replace copper wires. In biology, chemistry has changed the way scientists view life. Biochemists have found that all forms of life share many of the same molecules and molecular processes.

1.2   An experiment is an observation of natural phenomena carried out in a controlled manner so the results can be duplicated and rational conclusions obtained. A theory is a tested explanation of basic natural phenomena. They are related in that a theory is based on the results of many experiments and is fruitful in suggesting other new experiments. Also, an experiment can disprove a theory but never prove it absolutely. A hypothesis is a tentative explanation of some regularity of nature.

1.3   Rosenberg conducted controlled experiments and noted a basic relationship that could be stated as a hypothesis; that is, that certain platinum compounds inhibit cell division. This led him to do new experiments on the anti-cancer activity of these compounds.

1.4   Matter is the general term for the material things around us. It is whatever occupies space and can be perceived by our senses. Mass is the quantity of matter in a material. The difference between mass and weight is that mass remains the same wherever it is measured, but weight is proportional to the mass of the object divided by the square of the distance between the center of mass of the object and that of the earth.

1.5   The law of conservation of mass states that the total mass remains constant during a chemical change (chemical reaction). To demonstrate this law, place a sample of wood in a sealed vessel with air, and weigh it. Heat the vessel to burn the wood, and weigh the vessel after the experiment. The weight before and after the experiment should be the same.

1.6   Mercury metal, which is a liquid, reacts with oxygen gas to form solid mercury(II) oxide. The color changes from that of metallic mercury (silvery) to a color that varies from red to yellow depending on the particle size of the oxide.

1.7   An example of an element: sodium; of a compound: sodium chloride, or table salt; of a heterogeneous mixture: salt and sugar; of a homogeneous mixture: sodium chloride dissolved in water to form a solution.

1.8   The precision refers to the closeness of the set of values obtained from identical measurements of a quantity. The number of digits reported for the value of a measured or calculated quantity (significant figures) indicates the precision of the value.

1.9   An exact number is a number that arises when you count items or sometimes when you define a unit. For example, a foot is defined to be 12 inches. A measured number is the result of a comparison of a physical quantity with a fixed standard of measurement. For example, a steel rod measures 9.12 centimeters or 9.12 times the standard centimeter unit of measurement.

1.10   For a given unit, the SI system uses prefixes to obtain units of different sizes

1.11   An absolute temperature scale is a scale in which the lowest temperature that can be attained theoretically is zero. Degrees Celsius and kelvins have equal size units and are related by the formula

$$t_C = (T_K - 273.15\ K) \times \frac{1°C}{1\ K}$$

1.12   Units should be carried along because first, the units for the answers will come out in the calculations, and second, if you make an error in arranging factors in the calculation, this will become apparent because the final units will be nonsense.

■ Solutions to Practice Problems

Note on significant figures: If the final answer to a solution needs to be rounded off, it is given first with one nonsignificant figure, and the last significant figure is underlined. The final answer is then rounded to the correct number of significant figures. In multiple-step problems, intermediate answers are given with at least one nonsignificant figure; however, only the final answer has been rounded off.

1.19   By the law of conservation of mass:

Mass of zinc + mass of sulfur = mass of zinc sulfide

Rearranging and plugging in gives

Mass of zinc sulfide = 65.4 g + 32.1 g = 97.5 g

For the second part, let x = mass of zinc sulfide that could be produced. By the law of conservation of mass:

20.0 g + mass of sulfur = x

Write a proportion that relates the mass of zinc reacted to the mass of zinc sulfide formed, which should be the same for both cases.

$$\frac{\text{mass zinc}}{\text{mass zinc sulfide}} = \frac{65.4 \text{ g}}{97.5 \text{ g}} = \frac{20.0 \text{ g}}{x}$$

Solving gives x = 29.8̲1 g = 29.8 g

1.21 a. Solid   b. Liquid   c. Solid

1.23 a. Physical property   b. Chemical property
     c. Physical property   d. Physical property

1.25   Physical properties: (1) Iodine is solid; (2) the solid has lustrous blue-black crystals; (3) the crystals vaporize readily to a violet-colored gas.

Chemical properties: (1) Iodine combines with many metals such as with aluminum to give aluminum iodide.

1.27 a. Physical process   b. Chemical reaction
     c. Physical reaction   d. Chemical process

1.29 a. A pure substance with three phases present; solid, liquid, and gas.

    b. A mixture with two phases present; solid and liquid.

    c. A mixture with two phases present; solid and solid.

1.31 a. four   b. three   c. four   d. five

1.33   40,000 km = 4.0×10^4 km

1.35 a. $0.71 + 81.8 = 82.\underline{5}1 = 82.5$

    b. $134 \times 0.00435 + 107 = 0.58\underline{2}9 + 107 = 10\underline{7}.5829 = 108$

    c. $(847.89 - 847.73) \times 14673 = 0.1\underline{6} \times 14673 = 2\underline{3}47 = 2.3 \times 10^3$

1.37   The volume of the first sphere is

$$V_1 = (4/3)\pi r^3 = (4/3) \times 3.1416 \times (5.10 \text{ cm})^3 = 55\underline{5}.64 \text{ cm}^3$$

The volume of the second sphere is

$$V_2 = (4/3)\pi r^3 = (4/3) \times 3.1416 \times (5.00 \text{ cm})^3 = 52\underline{3}.60 \text{ cm}^3$$

The difference in volume is

$$V_1 - V_2 = 55\underline{5}.64 \text{ cm}^3 - 52\underline{3}.60 \text{ cm}^3 = 3\underline{2}.04 \text{ cm}^3 = 32 \text{ cm}^3$$

1.39 a. 5.89×10^{-9} s = 5.89 ns   b. 0.2010 m = 20.1 cm

1.41 a. 6.15 ps = 6.15×10^{-12} s   b. 3.781 μm = 3.781×10^{-6} m

1.43 a. $t_C = \dfrac{5°C}{9°F} \times (t_F - 32°F) = \dfrac{5°C}{9°F} \times (68°F - 32°F) = 2\underline{0}.0°C = 20.°C$

    b. $t_F = (t_C \times \dfrac{9°F}{5°C}) + 32°F = (-70°C \times \dfrac{9°F}{5°C}) + 32°F = -9\underline{4}.0°F = -94°F$

1.45 $t_F = (t_C \times \dfrac{9°F}{5°C}) + 32°F = (-21.1°C \times \dfrac{9°F}{5°C}) + 32°F = -5.\underline{9}8°F = -6.0°F$

1.47   First, determine the density of the liquid.

$$d = \frac{m}{V} = \frac{6.74 \text{ g}}{8.5 \text{ mL}} = 0.7\underline{9}29 = 0.79 \text{ g/mL}$$

The density is closest to ethanol (0.789 g/cm^3).

1.49   The mass of platinum is obtained as follows.

$$\text{Mass} = d \times V = 21.4 \text{ g/cm}^3 \times 8.3 \text{ cm}^3 = 1\underline{7}7 \text{ g} = 1.8 \times 10^2 \text{ g}$$

1.51   Since $1 \text{ kg} = 10^3 \text{ g}$, and $1 \text{ mg} = 10^{-3} \text{ g}$, you can write

$$0.480 \text{ kg} \times \frac{10^3 \text{ g}}{1 \text{ kg}} \times \frac{1 \text{ mg}}{10^{-3} \text{ g}} = 4.80 \times 10^5 \text{ mg}$$

1.53 $3.58 \text{ short ton} \times \dfrac{2000 \text{ lb}}{1 \text{ short ton}} \times \dfrac{16 \text{ oz}}{1 \text{ lb}} \times \dfrac{1 \text{ g}}{0.03527 \text{ oz}}$

$$= 3.2\underline{4}8 \times 10^6 \text{ g} = 3.25 \times 10^6 \text{ g}$$

1.55 $2425 \text{ fathoms} \times \dfrac{6 \text{ ft}}{1 \text{ fathom}} \times \dfrac{12 \text{ in}}{1 \text{ ft}} \times \dfrac{2.54 \times 10^{-2} \text{ m}}{1 \text{ in}} = 443\underline{4}.8 \text{ m} = 4435 \text{ m}$

■ Solutions to General Problems

1.57   From the law of conservation of mass,

Mass of aluminum + mass of iron(III) oxide

= mass of iron + mass of aluminum oxide + mass of unreacted iron(III) oxide

$5.40 \text{ g} + 18.50 \text{ g} = 11.17 \text{ g} + 10.20 \text{ g} +$ mass of iron(III) oxide unreacted

mass of iron(III) oxide unreacted = $5.40 \text{ g} + 18.50 \text{ g} - 11.17 \text{ g} - 10.20 \text{ g}$

$$= 2.5\underline{3} \text{ g}$$

Thus the mass of unreacted iron(III) oxide is 2.53 g.

1.59 a. Bromine   b. Phosphorus   c. Gold   d. Carbon (as graphite)

1.61   Compounds always contain the same proportions of the elements by mass. Thus, if we let X be the proportion of iron in a sample, we can calculate the proportion of iron in each sample as follows.

Sample A: $X = \dfrac{\text{mass of iron}}{\text{mass of sample}} = \dfrac{1.094\ g}{1.518\ g} = 0.7206\underline{8} = 0.7207$

Sample B: $X = \dfrac{\text{mass of iron}}{\text{mass of sample}} = \dfrac{1.449\ g}{2.056\ g} = 0.7047\underline{6} = 0.7048$

Sample C: $X = \dfrac{\text{mass of iron}}{\text{mass of sample}} = \dfrac{1.335\ g}{1.873\ g} = 0.7127\underline{6} = 0.7128$

Because each sample has a different proportion of iron by mass, the material is not a compound.

1.63 $V = (\text{edge})^3 = (19.4\ cm)^3 = 7.3\underline{0}1 \times 10^3\ cm^3 = 7.30 \times 10^3\ cm^3$

1.65 $V = LWH = 47.8\ in \times 12.5\ in \times 19.5\ in \times \dfrac{1\ gal}{231\ in^3} = 50.\underline{4}3\ gal = 50.4\ gal$

1.67 a. $\dfrac{56.1 - 51.1}{6.58} = 7.59 \times 10^{-1} = 7.6 \times 10^{-1}$

   b. $\dfrac{56.1 + 51.1}{6.58} = 1.6\underline{2}9 \times 10^1 = 1.63 \times 10^1$

   c. $(9.1 + 8.6) \times 26.91 = 4.7\underline{6}3 \times 10^2 = 4.76 \times 10^2$

   d. $0.0065 \times 3.21 + 0.0911 = 1.1\underline{1}9 \times 10^{-1} = 1.12 \times 10^{-1}$

1.69 a. $1.07 \times 10^{-12}\ s$ b. $5.8 \times 10^{-6}\ m$ c. $3.19 \times 10^{-7}\ m$ d. $1.53 \times 10^7\ s$

1.71 $t_F = (t_C \times \dfrac{9°F}{5°C}) + 32°F = (825°C \times \dfrac{9°F}{5°C}) + 32°F = 15\underline{1}7°F = 1.52 \times 10^3\ °F$

1.73   The temperature in degrees Celsius is

$$t_C = \frac{5°C}{9°F} \times (t_F - 32°F) = \frac{5°C}{9°F} \times (787°F - 32°F) = 419.44°C = 419°C$$

The temperature in kelvin units is

$$T_K = (t_C \times \frac{1\,K}{1°C}) + 273.15\,K = (419.44°C \times \frac{1\,K}{1°C}) + 273.15\,K$$

$$= 692.59 = 693\,K$$

1.75   Density $= \frac{1.74\,g}{1\,cm^3} \times \frac{1\,kg}{10^3\,g} \times \left(\frac{1\,cm}{10^{-2}\,m}\right)^3 = 1.74 \times 10^3\,kg/m^3$

1.77   First determine the density of the liquid sample.

$$Density = \frac{mass}{volume} = \frac{22.3\,g}{15.0\,mL} = 1.486\,g/mL = 1.49\,g/mL = 1.49\,g/cm^3$$

This density is closest to that of chloroform (1.489 g/cm³), so the unknown liquid is chloroform.

1.79   Volume $= \frac{mass}{density} = \frac{35.00\,g}{1.053\,g/mL} = 33.238\,mL = 33.24\,mL$

1.81 a. 5.91 kg $\times \frac{10^3\,g}{1\,kg} \times \frac{1\,mg}{10^{-3}\,g} = 5.91 \times 10^6\,mg$

b. 753 mg $\times \frac{10^{-3}\,g}{1\,mg} \times \frac{1\,\mu g}{10^{-6}\,g} = 7.53 \times 10^5\,\mu g$

c. 90.1 MHz $\times \frac{10^6\,Hz}{1\,MHz} \times \frac{1\,kHz}{10^3\,Hz} = 9.01 \times 10^4\,kHz$

1.83   Mass $= 275$ carats $\times \frac{200\,mg}{1\,carat} \times \frac{10^{-3}\,g}{1\,mg} = 55.00\,g = 55.0\,g$

■ Solutions to Cumulative-Skills Problems

1.85 First, calculate the volume of the steel sphere.

$$V = (4/3)\, \pi r^3 = (4/3) \times 3.1416 \times (1.58 \text{ in})^3 \times \left(\frac{2.54 \text{ cm}}{1 \text{ in}}\right)^3 = 27\underline{0}.7 \text{ cm}^3$$

Next, determine the mass of the sphere using the density.

Mass = density $\times$ volume = 7.88 g/cm^3 $\times$ 27$\underline{0}$.7 cm^3

$$= 21\underline{3}3 \text{ g} = 2.13 \times 10^3 \text{ g}$$

1.87 The area of the ice is 840,000 mi^2 - 132,000 mi^2 = 708,000 mi^2. Now determine the volume of this ice.

Volume = area $\times$ thickness

$$= 708{,}000 \text{ mi}^2 \times 5000 \text{ ft} \times \left(\frac{5280 \text{ ft}}{1 \text{ mi}}\right)^2 \times \left(\frac{12 \text{ in}}{1 \text{ ft}}\right)^3 \times \left(\frac{2.54 \text{ cm}}{1 \text{ in}}\right)^3$$

$$= 2.794 \times 10^{21} \text{ cm}^3$$

Now use the density to determine the mass of the ice.

mass = density $\times$ volume = 0.917 g/cm^3 $\times$ 2.794 $\times$ 10^{21} cm^3 = 2.$\underline{5}$6 $\times$ 10^{21} g

$$= 2.6 \times 10^{21} \text{ g}$$

1.89 Let x = mass of ethanol and y = mass of water. Then use the total mass to write x + y = 49.6 g, or y = 49.6 g - x. So the mass of water is 49.6 g - x. Next,

Total volume = volume of ethanol + volume of water

Since the volume is equal to the mass divided by density, you can write

$$\text{Total volume} = \frac{\text{mass of ethanol}}{\text{density of ethanol}} + \frac{\text{mass of water}}{\text{density of water}}$$

(continued)

Substitute in the known and unknown values to get an equation for x.

$$54.2 \text{ cm}^3 = \frac{x}{0.789 \text{ g/ cm}^3} + \frac{49.6 \text{ g - x}}{0.998 \text{ g/ cm}^3}$$

Multiply both sides of this equation by (0.789)(0.998). Also multiply both sides by g/cm³ to simplify the units. This gives the following equation to solve for x.

(0.789)(0.998)(54.2) g = (0.998) x + (0.789)(49.6 g - x)

42.678 g = 0.998 x + 39.134 g - 0.789 x

0.209 x = 3.544 g

x = mass of ethanol = 16.95 g

The percentage of ethanol (by mass) in the solution can now be calculated.

$$\% \text{ (mass)} = \frac{\text{mass of ethanol}}{\text{mass of solution}} \times 100\% = \frac{16.95 \text{ g}}{49.6 \text{ g}} \times 100\% = 34.1\% = 34\%$$

To determine the proof, you must first find the percentage by volume of ethanol in the solution. The volume of ethanol is obtained using the mass and the density.

$$\text{Volume} = \frac{\text{mass of ethanol}}{\text{density of ethanol}} = \frac{16.95 \text{ g}}{0.789 \text{ g/ cm}^3} = 21.48 \text{ cm}^3$$

The percentage of ethanol (by volume) in the solution can now be calculated.

$$\% \text{ (volume)} = \frac{\text{volume of ethanol}}{\text{volume of solution}} \times 100\% = \frac{21.48 \text{ cm}^3}{54.2 \text{ cm}^3} \times 100\% = 39.63\%$$

The proof can now be calculated.

Proof = 2 x % (volume) = 2 x 39.63 = 79.27 = 79 proof

2. ATOMS, MOLECULES, AND IONS

■ Answers to Review Questions

2.1   Atomic theory is an explanation of the structure of matter in terms of different combinations of very small particles called atoms. Since compounds are composed of atoms of two or more elements, there is no limit to the number of ways in which the elements can be combined. Each compound has its own unique properties. A chemical reaction consists of the rearrangement of the atoms present in the reacting substances to give new chemical combinations present in the substances formed by the reaction.

2.2   Divide each amount of chlorine, 1.270 g and 1.904 g, by the lower amount, 1.270 g. This gives 1.000 and 1.499, respectively. Convert these to whole numbers by multiplying by 2, giving 2.000 and 2.998. The ratio of these amounts of chlorine is essentially 2:3. This is consistent with the law of multiple proportions because, for a fixed mass of iron (one gram), the masses of chlorine in the other two compounds are in the ratio of small whole numbers.

2.3   A cathode-ray tube consists of a negative electrode, or cathode, and a positive electrode, or anode, in an evacuated tube. Cathode rays travel from the cathode to the anode when a high voltage is turned on. Some of the rays pass through the hole in the anode to form a beam, which is then bent toward positively charged electric plates in the tube. This implies that a cathode ray consists of a beam of negatively charged particles (or electrons) and that electrons are constituents of all matter.

2.4   Millikan performed a series of experiments in which he obtained the charge on the electron by observing how a charged drop of oil falls in the presence and in the absence of an electric field. An atomizer introduces a fine mist of oil drops into the top chamber (Figure 2.4). Several drops happen to fall through a small hole into the lower chamber where the experimenter follows the motion of one drop with a microscope. Some of these drops have picked up one or more electrons as a result of friction in the atomizer and have become negatively charged. A negatively charged drop will be attracted upwards when the experimenter turns on a current to the electric plates. The drop's upward speed (obtained by timing its rise) is related to its mass-to-charge ratio, from which you can calculate the charge on the electron.

2.5   The nuclear model of the atom is based on experiments of Geiger, Marsden, and Rutherford. Rutherford stated that most of the mass of an atom is concentrated in a positively charged center called the nucleus around which negatively charged electrons move. The nucleus, although containing most of the mass, occupies only a very small portion of the space of the atom. Most of the alpha particles passed through the metal atoms of the foil undeflected by the lightweight electrons. When an alpha particle does happen to hit a metal-atom nucleus, it is scattered at a wide angle because it is deflected by the massive, positively charged nucleus (Figure 2.6).

2.6   Protons (hydrogen nuclei) were discovered as products of experiments involving the collision of alpha particles with nitrogen atoms that resulted in a proton being knocked out of the nitrogen nucleus. Neutrons were discovered as the radiation product of collisions of alpha particles with beryllium atoms. The resulting radiation was discovered to consist of particles having a mass approximately equal to that of a proton and having no charge (neutral).

2.7   A mass spectrometer works by allowing a gas, such as neon, to pass through an inlet tube into a chamber where the atoms collide with electrons from an electron beam (cathode rays). The force of a collision can knock an electron from an atom. The positive neon atoms produced this way are drawn toward a negative grid, and some of them pass through slits to form a beam of positive electricity. This beam then travels through a magnetic field (from a magnet whose poles are above and below the positive beam). The magnetic field deflects the positively charged atoms in the positive beam according to their mass-to-charge ratios, and they then travel to a detector at the end of the tube. The beam of neon is split into three beams corresponding to the three different isotopes of neon (Figure 2.8). You obtain the mass of each neon isotope from the magnitude of deflection of its positively charged atoms. You can also obtain the relative number of atoms for the various masses (fractional abundances). The mass spectrum gives us all the information needed to calculate the atomic weight.

2.8   The atomic weight of an element is the average atomic mass for the naturally occurring element expressed in atomic mass units. The atomic weight would be different elsewhere in the universe if the percentages of isotopes in the element were different from those on earth.

2.9   The element in Group IVA and Period 5 is tin (atomic number 50).

2.10   A metal is a substance or mixture that has characteristic luster, or shine, and is generally a good conductor of heat and electricity.

2.11   The formula for ethane is C_2H_6.

2.12   Organic molecules contain carbon combined with other elements such as hydrogen, oxygen, and nitrogen. An inorganic molecule is composed of elements other than carbon. Some inorganic molecules that contain carbon are carbon monoxide (CO), carbon dioxide (CO_2), carbonates, and cyanides.

2.13   An ionic binary compound: NaCl; a molecular binary compound: H_2O.

2.14 a. The elements are represented by B, F, and I.

   b. The compounds are represented by A, E, and G.

   c. The mixtures are represented by C, D, and H.

   d. The ionic solid is represented by A.

   e. The gas made up of an element and a compound is represented by C.

   f. The mixtures of elements are represented by D and H.

   g. The solid element is represented by F.

   h. The solids are represented by A and F.

   i. The liquids are represented by E, H, and I.

2.15   In the Stock system, CuCl is called copper(I) chloride, and $CuCl_2$ is called copper(II) chloride.

   An advantage of the Stock system is that more than two different ions of the same metal can be named with this system. In the former (older) system, a new suffix other than -ic and -ous must be established and/or memorized.

2.16   A balanced chemical equation has the number of atoms of each element equal on both sides of the arrow. The coefficients are the smallest possible whole numbers.

■ Solutions to Practice Problems

Note on significant figures: If the final answer to a solution needs to be rounded off, it is given first with one nonsignificant figure, and the last significant figure is underlined. The final answer is then rounded to the correct number of significant figures. In multiple-step problems, intermediate answers are given with at least one nonsignificant figure; however, only the final answer has been rounded off.

2.21 a. Sodium   b. Zinc   c. Silver   d. Magnesium

2.23   The isotope of atom A is the atom with 17 protons, atom D; the atom that has the same mass number as atom A (36) is atom C.

2.25   Each isotope of copper (atomic number 29) has 29 protons. The neutral atoms will also each have 29 electrons. The number of neutrons for Cu-63 is 63 – 29 = 34 neutrons. The number of neutrons for Cu-65 is 65 – 29 = 36 neutrons.

2.27   Since the atomic ratio of nitrogen to hydrogen is 1:3, divide the mass of N by one-third of the mass of hydrogen to find the relative mass of N.

$$\frac{\text{Atomic mass of N}}{\text{Atomic mass of H}} = \frac{7.933\ g}{1/3 \times 1.712\ g} = \frac{13.901\ g\ N}{1\ g\ H} = \frac{13.90}{1}$$

2.29   Multiply each isotopic mass by its fractional abundance, then sum:

$$
\begin{aligned}
38.964 \times 0.9326 &= 36.3378 \\
39.964 \times 1.00 \times 10^{-4} &= 0.0039964 \\
40.962 \times 0.0673 &= \underline{2.75674} \\
&= 39.09853 = 39.099\ \text{amu}
\end{aligned}
$$

The atomic weight of this element is 39.099 amu. The element is potassium (K).

2.31   According to the picture, there are twenty atoms, five of which are brown and fifteen of which are green. Using the isotopic masses in the problem, the atomic mass of element X is

$$\frac{5}{20}\,(23.02\ \text{amu}) + \frac{15}{20}\,(25.147\ \text{amu}) = 5.755 + 18.8602 = 24.615 = 24.62\ \text{amu}$$

2.33 a.   C: Group IVA, Period 2; nonmetal   b.   Se: Group VIA, Period 4; metal

c.   Cr: Group VIB, Period 4; metal   d.   Be: Group IIA, Period 2; metal

e.   B: Group IIIA, Period 2; metalloid

2.35   Examples are:

a.   S (sulfur)   b.   Na (sodium)

c.   Fe (iron)   d.   U (uranium)

2.37   The number of nitrogen atoms in the 1.50 g sample of N_2O is

$$4.10 \times 10^{22} \text{ } N_2O \text{ molecules } \times \frac{2 \text{ N atoms}}{1 \text{ } N_2O \text{ molecule}} = 8.20 \times 10^{22} \text{ N atoms}$$

The number of nitrogen atoms in 1.00 g of N_2O is

$$1.00 \text{ g } \times \frac{8.20 \times 10^{22} \text{ N atoms}}{3.00 \text{ g } N_2O} = 2.7\underline{3}3 \times 10^{22} \text{ N atoms} = 2.73 \times 10^{22} \text{ N atoms}$$

2.39 a. PCl_5 b. NO_2 c. $C_3H_6O_2$

2.41 $\dfrac{2 \text{ Fe atom}}{1 \text{ } Fe_2(SO_4)_3 \text{ unit}} \times \dfrac{1 \text{ } Fe_2(SO_4)_3 \text{ unit}}{3 \text{ } SO_4^{2-} \text{ ions}} \times \dfrac{1 \text{ } SO_4^{2-} \text{ ion}}{4 \text{ O atoms}} = \dfrac{2 \text{ Fe atom}}{12 \text{ O atoms}} = \dfrac{1}{6}$

Thus, the ratio of iron atoms to oxygen atoms is one Fe atom to six O atoms.

2.43 a. $Fe(CN)_3$ b. K_2SO_4 c. Li_3N d. Ca_3P_2

2.45 a. K_2SO_3: potassium sulfite (Group IA forms only 1+ cations).

    b. CaO: calcium oxide (Group IIA forms only 2+ cations).

    c. CuCl: copper(I) chloride (Group IB forms 1+ and 2+ cations).

    d. Cr_2O_3: chromium(III) oxide (Group VIB forms numerous oxidation states).

2.47 a. Lead(II) chromate: $PbCrO_4$ (Chromate is in Table 2.6).

    b. Sodium hydrogen carbonate: $NaHCO_3$ (The HCO_3^- ion is in Table 2.6).

    c. Cesium sulfide: Cs_2S (Group 1A ions form 1+ cations).

    d. Iron(II) phosphate: $Fe_3(PO_4)_2$ (The phosphate ion = 3-; two PO_4^{3-} must be used to balance the three Fe^{2+}).

2.49 a. Selenium trioxide

    b. Disulfide dichloride

    c. Carbon monoxide

2.51 a. Bromic acid: $HBrO_3$ b. Nitrous acid: HNO_2

    c. Sulfurous acid: H_2SO_3 d. Arsenic acid: H_3AsO_4

2.53 $Na_2SO_4 \cdot 10H_2O$ is sodium sulfate decahydrate.

2.55   Iron(II) sulfate heptahydrate is $FeSO_4 \cdot 7H_2O$.

2.57 $Pb(NO_3)_2 \times \dfrac{6 \text{ O atoms}}{1 \text{ } Pb(NO_3)_2 \text{ unit}} + K_2CO_3 \times \dfrac{3 \text{ O atoms}}{1 \text{ } K_2CO_3 \text{ unit}} = 9 \text{ O atoms}$

2.59 a. Balance: $Sn + NaOH \rightarrow Na_2SnO_2 + H_2$

If Na is balanced first by writing a two in front of NaOH, the entire equation is balanced.

$Sn + 2NaOH \rightarrow Na_2SnO_2 + H_2$

b. Balance: $Mg + Fe_3O_4 \rightarrow MgO + Fe$

First balance Fe and O (they appear once on each side) by writing a four in front of MgO and a three in front of Fe:

$Mg + Fe_3O_4 \rightarrow 4MgO + 3Fe$

Now balance Mg against the four Mg's on the right:

$4Mg + Fe_3O_4 \rightarrow 4MgO + 3Fe$

c. Balance: $C_2H_5OH + O_2 \rightarrow CO_2 + H_2O$

First balance C and H (they appear once on each side) by writing a three in front of H_2O and a two in front of CO_2:

$C_2H_5OH + O_2 \rightarrow 2CO_2 + 3H_2O$

Finally, balance O:

$C_2H_5OH + 3O_2 \rightarrow 2CO_2 + 3H_2O$

d. Balance: $P_4O_6 + H_2O \rightarrow H_3PO_3$

First balance P (it appears once on each side) by writing a four in front of H_3PO_3:

$P_4O_6 + H_2O \rightarrow 4H_3PO_3$

Finally, balance H by writing a six in front of H_2O; this balances the entire equation:

$P_4O_6 + 6H_2O \rightarrow 4H_3PO_3$

(continued)

e. Balance: $PCl_5 + H_2O \rightarrow H_3PO_4 + HCl$

First balance Cl (it appears once on each side) by writing a five in front of HCl:

$$PCl_5 + H_2O \rightarrow H_3PO_4 + 5HCl$$

Finally, balance H by writing a four in front of H_2O; this balances the entire equation:

$$PCl_5 + 4H_2O \rightarrow H_3PO_4 + 5HCl$$

2.61　Balance: $Ca_3(PO_4)_2(s) + H_2SO_4(aq) \rightarrow CaSO_4(s) + H_3PO_4(aq)$

Balance Ca first with a three in front of $CaSO_4$:

$$Ca_3(PO_4)_2(s) + H_2SO_4(aq) \rightarrow 3CaSO_4(s) + H_3PO_4(aq)$$

Next, balance the P with a two in front of H_3PO_4:

$$Ca_3(PO_4)_2(s) + H_2SO_4(aq) \rightarrow 3CaSO_4(s) + 2H_3PO_4(aq)$$

Finally, balance the S with a three in front of H_2SO_4; this balances the equation:

$$Ca_3(PO_4)_2(s) + 3H_2SO_4(aq) \rightarrow 3CaSO_4(s) + 2H_3PO_4(aq)$$

■ Solutions to General Problems

2.63　Calculate the ratio of oxygen for one g (fixed amount) of nitrogen in both compounds:

A: $\dfrac{2.755 \text{ g O}}{1.206 \text{ g N}} = \dfrac{2.2844 \text{ g O}}{1 \text{ g N}}$　　B: $\dfrac{4.714 \text{ g O}}{1.651 \text{ g N}} = \dfrac{2.8552 \text{ g O}}{1 \text{ g N}}$

Next, find the ratio of oxygen per gram of nitrogen for the two compounds.

$$\frac{\text{g O in B/1 g N}}{\text{g O in A/1 g N}} = \frac{2.8552 \text{ g O}}{2.2844 \text{ g O}} = \frac{1.2498 \text{ g O}}{1 \text{ g O}}$$

B contains 1.25 times as many O atoms as A does (there are five O's in B for every four O's in A).

2.65　For the Eu atom to be neutral, the number of electrons must equal the number of protons, so a neutral europium atom has 63 electrons. The +3 charge on the Eu^{3+} indicates there are three more protons than electrons, so the number of electrons = 63 - 3 = 60.

2.67 The number of protons = mass number - number of neutrons = 81 - 46 = 34. The element with Z = 34 is selenium (Se).

The ionic charge = number of protons - number of electrons = 34 - 36 = -2.

Symbol: $^{80}_{34}Se^{2-}$.

2.69 The sum of the fractional abundances must equal one. Thus, the abundance of one isotope can be expressed in terms of the other. Let y equal the fractional abundance of europium-1. Then the fractional abundance of europium-2 equals (1 - y). We can write one equation in one unknown:

Atomic weight = 151.96 = 150.92y + 152.92(1 - y)

$$151.96 = 152.92 - 2.00y$$

$$y = \frac{152.92 - 151.96}{2.00} = 0.48\underline{0}0$$

The fractional abundance of europium-1 (Eu-151) = 0.48$\underline{0}$0 = 0.480.

The fractional abundance of europium-2 (Eu-153) = 1 - 0.48$\underline{0}$0 = 0.52$\underline{0}$0 = 0.520.

2.71 a. Mercury, Hg   b. Hydrogen, H   c. Niobium, Nb   d. Astatine, At

2.73 All possible ionic compounds: Na_2CO_3, $NaCl$, $MgCO_3$, and $MgCl_2$.

2.75 a. Hg_2S [Mercury(I) exists as the polyatomic Hg_2^{2+} ion (Table 2.6).]

   b. $Co_2(SO_3)_3$ c. $(NH_4)_2Cr_2O_7$ d. Al_2S_3

2.77 a. Arsenic tribromide   b. hydrogen telluride (dihydrogen telluride)
   c. Diphosphorus pent(a)oxide   d. Silicon dioxide

2.79 a. Balance the C and H first:

   $$C_3H_8 + O_2 \rightarrow 3CO_2 + 4H_2O$$

   Finally, balance the O's.

   $$C_3H_8 + 5O_2 \rightarrow 3CO_2 + 4H_2O$$

(continued)

b. Balance the P first:

$$P_4O_6 + H_2O \rightarrow 4H_3PO_3$$

Then balance the O (or H), which also gives the H (or O) balance:

$$P_4O_6 + 6H_2O \rightarrow 4H_3PO_4$$

c. Balancing the O first is the simplest approach. (Starting with K and Cl and then O will cause the initial coefficient for $KClO_3$ to be changed in balancing O last.)

$$4KClO_3 \rightarrow KCl + 3KClO_4$$

d. Balance the N first:

$$(NH_4)_2SO_4 + NaOH \rightarrow 2NH_3 + H_2O + Na_2SO_4$$

Then balance the Na, followed by O; this also balances the H:

$$(NH_4)_2SO_4 + 2NaOH \rightarrow 2NH_3 + 2H_2O + Na_2SO_4$$

2.81 Let: x = number of protons. Then $1.21x$ is the number of neutrons. Since the mass number is 62, you get

$$62 = x + 1.21x = 2.21x$$

Thus, x = 2̲8̲.054, or 28. The element is nickel (Ni). Since the ion has a +2 charge, there are 26 electrons.

■ Solutions to Cumulative-Skills Problems

2.83 $NiSO_4{\cdot}7H_2O(s) \rightarrow NiSO_4{\cdot}6H_2O(s) + H_2O(g)$

[8.753] = [8.192 g + (8.753 - 8.192 = 0.561 g)]

The 8.192 g of $NiSO_4{\cdot}6H_2O$ must contain 6×0.561 = 3.366 g H_2O.

Mass of anhydrous $NiSO_4$ = 8.192 g $NiSO_4{\cdot}6H_2O$ - 3.366 g $6H_2O$ = 4.826 g

2.85   Mass of O $= 0.6015$ L $\times \dfrac{1.330 \text{ g O}}{1 \text{ L}} = 0.799\underline{9}95$ g

15.9994 amu O $\times \dfrac{3.177 \text{ g X}}{0.799\underline{9}95 \text{ g O}} = 63.5\underline{3}8$ amu X $= 63.54$ amu

The atomic weight of X is 63.54 amu; X is copper.

3. CALCULATIONS WITH CHEMICAL FORMULAS AND EQUATIONS

■ Answers to Review Questions

3.1   The molecular weight is the sum of the atomic weights of all the atoms in a molecule of the substance whereas the formula weight is the sum of the atomic weights of all the atoms in one formula unit of the compound, whether the compound is molecular or not. A given substance could have both a molecular weight and a formula weight if it existed as discrete molecules.

3.2   To obtain the formula weight of a substance, sum up the atomic weights of all atoms in the formula.

3.3   A mole of N_2 contains Avogadro's number (6.02×10^{23}) of N_2 molecules and $2 \times 6.02 \times 10^{23}$ of N atoms. One mole of $Fe_2(SO_4)_3$ contains three moles of SO_4^{2-} ions; and it contains twelve moles of O atoms.

3.4   A sample of the compound of known mass is burned, and CO_2 and H_2O are obtained as products. Next, you relate the masses of CO_2 and H_2O to the masses of carbon and hydrogen. Then, you calculate the mass percentages of C and H. You find the mass percentage of O by subtracting the mass percentages of C and H from 100.

3.5   The empirical formula is obtained from the percentage composition by assuming for the purposes of the calculation a sample of 100 g of the substance. Then, the mass of each element in the sample equals the numerical value of the percentage. Convert the masses of the elements to moles of the elements using the atomic mass of each element. Divide the moles of each by the smallest number to obtain the smallest ratio of each atom. If necessary, find a whole-number factor to multiply these results by to obtain integers for the subscripts in the empirical formula.

3.6   The empirical formula is the formula of a substance written with the smallest integer (whole number) subscripts. Each of the subscripts in the formula $C_6H_{12}O_2$ can be divided by two, so the empirical formula of the compound is C_3H_6O.

3.7 The coefficients in a chemical equation can be interpreted directly in terms of molecules or moles. For the mass interpretation, you will need the molar masses of CH_4, O_2, CO_2, and H_2O, which are 16.0, 32.0, 44.0, and 18.0 g/mol, respectively. A summary of the three interpretations is given below the balanced equation:

CH_4	+	$2O_2$	→	CO_2	+	$2H_2O$
1 molecule	+	2 molecules	→	1 molecule	+	2 molecules
1 mole	+	2 moles	→	1 mole	+	2 moles
16.0 g	+	2 x 32.0 g	→	44.0 g	+	2 x 18.0 g

3.8 A chemical equation yields the mole ratio of a reactant to a second reactant or product. Once the mass of a reactant is converted to moles, this can be multiplied by the appropriate mole ratio to give the moles of a second reactant or product. Multiplying this number of moles by the appropriate molar mass gives mass. Thus, the masses of two different substances are related by a chemical equation.

3.9 The limiting reactant is the reactant that is entirely consumed when the reaction is complete. Because the reaction stops when the limiting reactant is used up, the moles of product are always determined by the starting number of moles of the limiting reactant.

3.10 Two examples are given in the book. The first involves making cheese sandwiches. Each sandwich requires two slices of bread and one slice of cheese. The limiting reactant is the cheese because some bread is left unused. The second example is assembling automobiles. Each auto requires one steering wheel, four tires, and other components. The limiting reactant is the tires, since they will run out first.

■ Solutions to Practice Problems

Note on significant figures: If the final answer to a solution needs to be rounded off, it is given first with one nonsignificant figure, and the last significant figure is underlined. The final answer is then rounded to the correct number of significant figures. In multiple-step problems, intermediate answers are given with at least one nonsignificant figure; however, only the final answer has been rounded off.

3.17 a SO_2 1 x AW of S   =   32.07 amu
                 2 x AW of O =   2 x 16.00   =   32.00 amu
                              MW of NO_2 =   64.07 amu = 64.1 amu (3 s.f.)

     b. PCl_3 1 x AW of P   =   30.97 amu
                 3 x AW of Cl =   3 x 35.45   =   106.35 amu
                              MW of PCl_3 =   137.32 amu = 137 amu (3 s.f.)

3.19 a. The atomic weight of K = 39.10 amu; thus, the molar mass = 39.10 g/mol. Because 1 mol of K atoms = 6.022 x 10^{23} K atoms, we calculate

$$\text{Mass of one K atom} = \frac{39.10 \text{ g/mol}}{6.022 \times 10^{23} \text{ atom/mol}} = 6.4929 \times 10^{-23} \text{ g}$$

b. The formula weight of $CHCl_3$ = [12.01 + (3 x 35.45) + 1.008] = 119.368 amu; thus, the molar mass = 119.37 g/mol. Because 1 mol of $CHCl_3$ molecules = 6.022 x 10^{23} $CHCl_3$ molecules, we calculate

$$\text{Mass of one } CHCl_3 \text{ molec} = \frac{119.368 \text{ g/mol}}{6.022 \times 10^{23} \text{ molec/mol}} = 1.9822 \times 10^{-22} \text{ g/molec}$$

3.21 From the table of atomic weights, we obtain the following molar masses for parts a through d: Na = 22.99 g/mol, S = 32.07 g/mol, C = 12.01 g/mol, H = 1.008 g/mol, Cl = 35.45 g/mol, and O = 16.00 g/mol.

a. 0.15 mol Na x $\frac{22.99 \text{ g}}{1 \text{ mol Na}}$ = 3.448 = 3.4 g Na

b. Using molar mass = 84.93 g/mol for CH_2Cl_2, we obtain

$$2.78 \text{ mol } CH_2Cl_2 \times \frac{84.93 \text{ g } CH_2Cl_2}{1 \text{ mol } CH_2Cl_2} = 236.1 = 236 \text{ g } CH_2Cl_2$$

3.23 From the table of atomic weights, we obtain the following rounded molar masses for parts a through d: C = 12.01 g/mol; Cl = 35.45 g/mol; H = 1.008 g/mol; Li = 6.94 g/mol; and O = 16.00 g/mol.

a. 2.86 g C x $\frac{1 \text{ mol C}}{12.01 \text{ g C}}$ = 0.2381 = 0.238 mol C

b. The molar mass of C_4H_{10} = (4 x 12.01) + (10 x 1.008) = 58.12 g C_4H_{10}/mol C_4H_{10}. The mass of C_4H_{10} is calculated as follows:

$$76 \text{ g } C_4H_{10} \times \frac{1 \text{ mol } C_4H_{10}}{58.12 \text{ g } C_4H_{10}} = 1.307 = 1.3 \text{ mol } C_4H_{10}$$

3.25 Calculate the formula weight of calcium sulfate: 40.08 amu + 32.07 amu + (4 x 16.00 amu) = 136.15 amu. Therefore, the molar mass of $CaSO_4$ is 136.15 g/mol. Use this to convert the mass of $CaSO_4$ to moles:

$$0.791 \text{ g CaSO}_4 \text{ x } \frac{1 \text{ mol CaSO}_4}{136.15 \text{ g CaSO}_4} = 5.8\underline{1}1 \text{ x } 10^{-3} = 5.81 \text{ x } 10^{-3} \text{ mol CaSO}_4$$

Calculate the molecular weight of water: (2 x 1.008 amu) + 16.00 amu = 18.02 amu. Therefore, the molar mass of H_2O = 18.02 g/mol. Use this to convert the rest of the sample to moles of water:

$$0.209 \text{ g H}_2\text{O x } \frac{1 \text{ mol H}_2\text{O}}{18.02 \text{ g H}_2\text{O}} = 1.1\underline{5}9 \text{ x } 10^{-2} = 1.16 \text{ x } 10^{-2} \text{ mol H}_2\text{O}$$

Because 0.01159 mol is about twice 0.005811 mol, both numbers of moles are consistent with the formula $CaSO_4 \cdot 2H_2O$.

3.27 The following rounded atomic weights are used: Br = 79.90 g/mol; N = 14.01 g/mol; H = 1.008 g/mol; O = 16.00 g/mol; Cr = 52.00 g/mol; and S = 32.07 g/mol. Also, Avogadro's number is 6.022 x 10^{23} atoms, so

a. No. Br atoms = 25.1 g Br_2 x $\dfrac{2 \text{ x } 6.022 \text{ x } 10^{23} \text{ atoms}}{(2 \text{ x } 79.90) \text{ g Br}_2}$ = 1.8$\underline{9}$1 x 10^{23} atoms

b. No. NH_3 molec. = 62 g NH_3 x $\dfrac{6.022 \text{ x } 10^{23} \text{ molec.}}{17.03 \text{ g NH}_3}$ = 2.$\underline{1}$9 x 10^{24} molec.

c. No.SO_4^{2-} ions = 21.4 g $Cr_2(SO_4)_3$ x $\dfrac{3 \text{ x } 6.022 \text{ x } 10^{23} \text{ ions}}{392.21 \text{ g Cr}_2(\text{SO}_4)_3}$ = 9.8$\underline{5}$7 x 10^{22} ions

3.29 Calculate the molecular weight of CCl_4: 12.01 amu + (4 x 35.45 amu) = 153.81 amu. Use this and Avogadro's number to express it as 153.81 g/N_A to calculate the number of molecules:

$$8.35 \text{ mg CCl}_4 \text{ x } \frac{1 \text{ g}}{1000 \text{ mg}} \text{ x } \frac{6.022 \text{ x } 10^{23} \text{ molec.}}{153.81 \text{ g CCl}_4} = 3.2\underline{6}9 \text{ x } 10^{19}$$

$$= 3.27 \text{ x } 10^{19} \text{ molec.}$$

3.31   Mass % phosphorus oxychloride $= \dfrac{\text{mass of } POCl_3 \text{ in sample}}{\text{mass of sample}} \times 100\%$

$$\% \ POCl_3 \ = \ \frac{1.72 \text{ mg}}{8.53 \text{ mg}} \ = \ 100\% \ = \ 20.\underline{1}6 \ = \ 20.2\%$$

3.33   Start with the definition for percentage nitrogen and rearrange this equation to find the mass of N in the fertilizer.

$$\text{Mass \% nitrogen} \ = \ \frac{\text{mass of N in fertilizer}}{\text{mass of fertilizer}} \times 100\%$$

$$\text{Mass N} \ = \ \frac{\text{mass \% N}}{100\%} \times \text{mass of fertilizer} \ = \ \frac{15.8\%}{100\%} \times 4.15 \text{ kg} \ = \ 0.65\underline{5}7$$

$$= \ 0.656 \text{ kg N}$$

3.35   In each part, the numerator consists of the mass of the element in one mole of the compound; the denominator is the mass of one mole of the compound. Use the atomic weights of N = 14.01 g/mol; O = 16.00 g/mol; Na = 22.99 g/mol; H = 1.008 g/mol; and P = 30.97 g/mol.

a. $\%N \ = \ \dfrac{\text{mass of N}}{\text{mass of } N_2O} \ = \ \dfrac{2 \times 14.01 \text{ g N}}{44.02 \text{ g } N_2O} \times 100\% \ = \ 63.652 \ = \ 63.7\%$

$\% \ O \ = \ 100.000\% \ - \ 63.652\%C \ = \ 36.347 \ = \ 36.4\%$

b. $\%Na \ = \ \dfrac{\text{mass of Na}}{\text{mass of } NaH_2PO_4} \ = \ \dfrac{22.99 \text{ g Na}}{119.98 \text{ g } NaH_2PO_4} \times 100\% \ = \ 19.161 \ = \ 19.2\%$

$\%H \ = \ \dfrac{\text{mass of H}}{\text{mass of } NaH_2PO_4} \ = \ \dfrac{2.016 \text{ g H}}{119.98 \text{ g } NaH_2PO_4} \times 100\% \ = \ 1.68\underline{0}2 \ = \ 1.68\%$

$\%P \ = \ \dfrac{\text{mass of P}}{\text{mass of } NaH_2PO_4} \ = \ \dfrac{30.97 \text{ g P}}{119.98 \text{ g } NaH_2PO_4} \times 100\% \ = \ 25.8\underline{1}2 \ = \ 25.8\%$

$\% \ O \ = \ 100.000\% \ - \ (19.161 \ + \ 1.6802 \ + \ 25.812) \ = \ 53.346 \ = \ 53.3\%$

3.37   The molecular model of toluene contains seven carbon atoms and eight hydrogen atoms, so the molecular formula of toluene is C_7H_8. The molar mass of toluene is 92.134 g/mol. The mass percentages are

$$\%C = \frac{\text{mass of C}}{\text{mass of } C_7H_8} = \frac{7 \times 12.01 \text{ g}}{92.134 \text{ g}} \times 100\% = 91.247 = 91.2\%$$

$$\%H = 100\% - 91.247 = 8.753 = 8.75\%$$

3.39   First calculate the mass of C in the glycol by multiplying the mass of CO_2 by the molar mass of C and the reciprocal of the molar mass of CO_2. Then calculate the mass of H in the glycol by multiplying the mass of H_2O by the molar mass of 2H and the reciprocal of the molar mass of H_2O. Then use the masses to calculate the mass percentages. Calculate O by difference.

$$9.06 \text{ mg } CO_2 \times \frac{1 \text{ mol } CO_2}{44.01 \text{ g } CO_2} \times \frac{12.01 \text{ g C}}{1 \text{ mol C}} = 2.4\underline{7}2 \text{ mg C}$$

$$5.58 \text{ mg } H_2O \times \frac{1 \text{ mol } H_2O}{18.02 \text{ g } H_2O} \times \frac{2 \text{ H}}{1 \text{ } H_2O} \times \frac{1.008 \text{ g H}}{1 \text{ mol H}} = 0.62\underline{4}3 \text{ mg H}$$

Mass O = 6.38 mg - (2.472 + 0.6243) = 3.2\underline{8}4 mg O

% C = (2.472 mg C/6.38 mg glycol) x 100% = 38.\underline{7}4 = 38.7%

% H = (0.6243 mg H/6.38 mg glycol) x 100% = 9.7\underline{8}5 = 9.79%

% O = (3.284 mg O/6.38 mg glycol) x 100% = 51.\underline{4}7 = 51.5%

3.41   Start by calculating the moles of Os and O; then divide each by the smaller number of moles to obtain integers for the empirical formula.

$$\text{Mol Os} = 2.16 \text{ g Os} \times \frac{1 \text{ mol Os}}{190.2 \text{ g Os}} = 0.0113\underline{6} \text{ mol (smaller number)}$$

$$\text{Mol O} = (2.89 - 2.16) \text{ g O} \times \frac{1 \text{ mol O}}{16.00 \text{ g O}} = 0.045\underline{6} \text{ mol}$$

Integer for Os = 0.01136 ÷ 0.01136 = 1.0\underline{0}0

Integer for O = 0.0456 ÷ 0.01136 = 4.\underline{0}1

Within experimental error, the empirical formula is OsO_4.

3.43 a. Assume for the calculation that you have 100.0 g; of this quantity, 88.83 g is C and 11.17 g is H. Now, convert these masses to moles:

$$88.83 \text{ g C} \times \frac{1 \text{ mol C}}{12.01 \text{ g C}} = 7.3\underline{9}6 \text{ mol C}$$

$$11.17 \text{ g H} \times \frac{1 \text{ mol H}}{1.008 \text{ g H}} = 11.081 \text{ mol H}$$

To obtain mole ratios divide each number of moles by the smallest number of moles, 7.3963 mol C. This gives

$$\text{For C:} \quad \frac{7.396 \text{ mol}}{7.396 \text{ mol}} = 1.00$$

$$\text{For H:} \quad \frac{11.081 \text{ mol}}{7.688 \text{ mol}} = 1.498$$

Finally, multiply both by two to eliminate the fractions. This gives C_2H_3 as the empirical formula for both compounds.

b. Obtain n, the number of empirical formula units in the molecule, by dividing the molecular weight of 54.08 amu and 108.16 amu, by the empirical formula weight of 27.044 amu:

$$\text{For 54.08: } n = \frac{54.08 \text{ amu}}{27.044 \text{ amu}} = 1.999, \text{ or 2}$$

$$\text{For 108.16: } n = \frac{108.16 \text{ amu}}{27.044 \text{ amu}} = 3.999, \text{ or 4}$$

The molecular formulas are: for 54.08, $(C_2H_3)_2$ or C_4H_6; and for 108.16, $(C_2H_3)_4$ or C_8H_{12}.

3.45 The formula weight corresponding to the empirical formula C_2H_6N may be found by adding the respective atomic weights.

$$\text{Formula weight} = (2 \times 12.01 \text{ amu}) + (6 \times 1.008 \text{ amu}) + 14.01 \text{ amu}$$
$$= 44.08 \text{ amu}$$

Dividing the molecular weight by the formula weight gives the number of times the C_2H_6N unit occurs in the molecule. Because the molecular weight is an average of 88.5 ([90 + 87] ÷ 2), this quotient is

$$88.5 \text{ amu} \div 44.1 \text{ amu} = 2.0\underline{0}6, \text{ or 2}$$

Therefore, the molecular formula is $(C_2H_6N)_2$, or $C_4H_{12}N_2$.

3.47 Assume a sample of 100.0 g of oxalic acid. By multiplying this by the percentage composition, we obtain 26.7 g C, 2.2 g H, and 71.1 g O. Convert each of these masses to moles by dividing by the molar mass.

$$\text{Mol C} = 26.7 \text{ g C} \times \frac{1 \text{ mol C}}{12.01 \text{ g C}} = 2.2\underline{2}3 \text{ mol}$$

$$\text{Mol H} = 2.2 \text{ g H} \times \frac{1 \text{ mol H}}{1.008 \text{ g H}} = 2.\underline{1}8 \text{ mol (smallest number)}$$

$$\text{Mol O} = 71.1 \text{ g O} \times \frac{1 \text{ mol O}}{16.00 \text{ g O}} = 4.4\underline{4}3 \text{ mol}$$

Now divide each number of moles by the smallest number to obtain the smallest set of integers for the empirical formula.

Integer for C = 2.223 ÷ 2.18 = 1.02, or 1

Integer for H = 2.18 ÷ 2.18 = 1.00, or 1

Integer for O = 4.443 ÷ 2.18 = 2.0$\underline{38}$, or 2

The empirical formula is thus CHO_2. The formula weight corresponding to this formula may be found by adding the respective atomic weights:

Formula weight = 12.01 amu + 1.008 amu + (2 × 16.00 amu) = 45.02 amu

Dividing the molecular weight by the formula weight gives the number of times the CHO_2 unit occurs in the molecule. Because the molecular weight is 90 amu, this quotient is

90 amu ÷ 45.02 amu = 2.$\underline{00}$, or 2

The molecular formula is thus $(CHO_2)_2$, or $C_2H_2O_4$.

3.49 $2C_3H_6$ + $9O_2$ → $6CO_2$ + $6H_2O$

2 molecule C_3H_6 + 9 molecules O_2 → 6 molecules CO_2 + 6 molecules H_2O

2 mole C_3H_6 + 9 moles O_2 → 6 moles CO_2 + 6 moles H_2O

2 × 42.078 g C_3H_6 + 9 × 32.00 g O_2 → 6 × 44.01 g CO_2 + 6 × 18.016 g H_2O

3.51   By inspecting the balanced equation, obtain a conversion factor of 3 mol O_2 to 2 mol Fe_2O_3. Multiply the given amount of 4.89 mol Fe_2O_3 by the conversion factor to obtain moles of O_2.

$$4.89 \text{ mol } Fe_2O_3 \times \frac{3 \text{ mol } O_2}{2 \text{ mol } Fe_2O_3} = 7.3\underline{3}5 = 7.34 \text{ mol } O_2$$

3.53 $WO_3 + 3H_2 \rightarrow W + 3H_2O$

1 mol of W is equivalent to 3 moles of H_2 (from equation).

1 mol of H_2 is equivalent to 2.016 g H_2 (from molecular weight of H_2).

1 mol of W is equivalent to 183.8 g W (from atomic weight of W).

4.81 kg of H_2 is equivalent to 4.81×10^3 g of H_2.

$$4.81 \times 10^3 \text{ g } H_2 \times \frac{1 \text{ mol } H_2}{2.016 \text{ g } H_2} \times \frac{1 \text{ mol } W}{3 \text{ mol } H_2} \times \frac{183.85 \text{ g } W}{1 \text{ mol } W} = 1.4\underline{6}2 \times 10^5$$

$$= 1.46 \times 10^5 \text{ g W}$$

3.55   Write the equation and set up the calculation below the equation (after calculating the two molecular weights):

$$CS_2 + 3Cl_2 \rightarrow CCl_4 + S_2Cl_2$$

$$54.9 \text{ g } Cl_2 \times \frac{1 \text{ mol } Cl_2}{70.90 \text{ g } Cl_2} \times \frac{1 \text{ mol } CS_2}{3 \text{ mol } Cl_2} \times \frac{76.15 \text{ g } CS_2}{1 \text{ mol } CS_2} \quad 19.\underline{6}55$$

$$= 19.7 \text{ g } CS_2$$

3.57   First determine whether KO_2 or H_2O is the limiting reactant by calculating the moles of O_2 that each would form if each were the limiting reactant. Identify the limiting reactant by the smaller number of moles of O_2 formed.

$$0.15 \text{ mol } H_2O \times \frac{3 \text{ mol } O_2}{2 \text{ mol } H_2O} = 0.2\underline{2}5 \text{ mol } O_2$$

$$0.25 \text{ mol } KO_2 \times \frac{3 \text{ mol } O_2}{4 \text{ mol } KO_2} = 0.1\underline{8}7 \text{ mol } O_2 \text{ (KO_2 is the limiting reactant)}$$

The moles of O_2 produced = 0.19 mol.

3.59 First determine whether CO or H_2 is the limiting reactant by calculating the moles of CH_3OH that each would form if each were the limiting reactant. Identify the limiting reactant by the smaller number of moles of CH_3OH formed. Use the molar mass of CH_3OH to calculate the mass of CH_3OH formed. Then calculate the mass of the unconsumed reactant.

$$CO + 2H_2 \rightarrow CH_3OH$$

$$8.25 \text{ g } H_2 \times \frac{1 \text{ mol } H_2}{2.016 \text{ g } H_2} \times \frac{1 \text{ mol } CH_3OH}{2 \text{ mol } H_2} = 2.0\underline{4}61 \text{ mol } CH_3OH$$

$$41.7 \text{ g CO} \times \frac{1 \text{ mol CO}}{28.01 \text{ g CO}} \times \frac{1 \text{ mol } CH_3OH}{1 \text{ mol CO}} = 1.4\underline{8}8 \text{ mol } CH_3OH$$

CO is the limiting reactant.

$$\text{Mass } CH_3OH \text{ formed } = 1.4\underline{8}8 \text{ mol } CH_3OH \times \frac{32.042 \text{ g } CH_3OH}{1 \text{ mol } CH_3OH}$$

$$= 47.\underline{7}0 = 47.7\text{g } CH_3OH$$

Hydrogen is left unconsumed at the end of the reaction. The mass of H_2 that reacts can be calculated from the moles of product obtained:

$$1.4\underline{8}8 \text{ mol } CH_3OH \times \frac{2 \text{ mol } H_2}{1 \text{ mol } CH_3OH} \times \frac{2.016 \text{ g } H_2}{1 \text{ mol } H_2} = 6.0\underline{0}2 \text{ g } H_2$$

The unreacted H_2 = 8.25 g total H_2 - 6.002 g reacted H_2 = 2.2$\underline{4}$7 = 2.25 g H_2.

3.61 First determine which of the two reactants is the limiting reactant by calculating the moles of aspirin that each would form if each were the limiting reactant. Identify the limiting reactant by the smallest number of moles of aspirin formed. Use the molar mass of aspirin to calculate the theoretical yield in grams of aspirin. Then calculate the percentage yield.

$$C_7H_6O_3 + C_4H_6O_3 \rightarrow C_9H_8O_4 + C_2H_4O_2$$

$$4.00 \text{ g } C_4H_6O_3 \times \frac{1 \text{ mol } C_4H_6O_3}{102.09 \text{ g } C_4H_6O_3} \times \frac{1 \text{ mol } C_9H_8O_4}{1 \text{ mol } C_4H_6O_3} = 0.039\underline{1}8 \text{ mol } C_9H_8O_4$$

$$2.00 \text{ g } C_7H_6O_3 \times \frac{1 \text{ mol } C_7H_6O_3}{138.12 \text{ g } C_7H_6O_3} \times \frac{1 \text{ mol } C_9H_8O_4}{1 \text{ mol } C_7H_6O_3} = 0.014\underline{4}8 \text{ mol } C_9H_8O_4$$

(continued)

Thus, $C_7H_6O_3$ is the limiting reactant. The theoretical yield of $C_9H_8O_4$ is

$$0.01448 \text{ mol } C_9H_8O_4 \times \frac{180.15 \text{ g } C_9H_8O_4}{1 \text{ mol } C_9H_8O_4} = 2.609 \text{ g } C_9H_8O_4$$

The percentage yield is

$$\% \text{ yield} = \frac{\text{actual yield}}{\text{theoretical yield}} \times 100\% = \frac{1.98 \text{ g}}{2.609 \text{ g}} \times 100\% = 75.89 = 75.9\%$$

■ Solutions to General Problems

3.63   For 1 mol of caffeine, there are 8 mol of C, 10 mol of H, 4 mol of N, and 2 mol of O. Convert these amounts to masses by multiplying by the respective molar masses:

8 mol C × 12.01 g C/1 mol C =		96.08 g C
10 mol H × 1.008 g H/1 mol H =		10.08 g H
4 mol N × 14.01 g N/1 mol N =		56.04 g N
2 mol O × 16.00 g O/1 mol O =		32.00 g O
1 mol of caffeine (total)	=	194.20 g (molar mass)

Each mass % is calculated by dividing the mass of the element by the molar mass of caffeine and multiplying by 100%: Mass % = (mass element ÷ mass caffeine) × 100%.

Mass % C = (96.08 g ÷ 194.20 g) × 100% = 49.5% (3 s.f.)

Mass % H = (10.08 g ÷ 194.20 g) × 100% = 5.19% (3 s.f.)

Mass % N = (56.04 g ÷ 194.20 g) × 100% = 28.9% (3 s.f.)

Mass % O = (32.00 g ÷ 194.20 g) × 100% = 16.5% (3 s.f.)

3.65   Assume a sample of 100.0 g of dichlorobenzene. By multiplying this by the percentage composition, we obtain 49.1 g C, 2.7 g of H, and 48.2 g of Cl. Convert each mass to moles by dividing by the molar mass:

$$49.1 \text{ g C} \times \frac{1 \text{ mol C}}{12.01 \text{ g C}} = 4.088 \text{ mol C}$$

$$2.7 \text{ g H} \times \frac{1 \text{ mol H}}{1.008 \text{ g H}} = 2.68 \text{ mol H}$$

(continued)

$$48.2 \text{ g Cl} \times \frac{1 \text{ mol Cl}}{35.45 \text{ g Cl}} = 1.3\underline{6}0 \text{ mol Cl}$$

Divide each number of moles by the smallest number to obtain the smallest set of integers for the empirical formula.

Integer for C = 4.088 mol ÷ 1.360 mol = 3.00, or 3

Integer for H = 2.68 mol ÷ 1.360 mol = 1.97, or 2

Integer for Cl = 1.360 mol ÷ 1.360 mol = 1.00, or 1

The empirical formula is thus C_3H_2Cl. Find the formula weight by adding the atomic weights:

Formula weight = (3 x 12.01 amu) + (2 x 1.008 amu) + 35.45 amu

$$= 73.4\underline{96} = 73.50 \text{ amu}$$

Divide the molecular weight by the formula weight to find the number of times the C_3H_2Cl unit occurs in the molecule. Because the molecular weight is 147 amu, this quotient is

147 amu ÷ 73.50 amu = 2.00, or 2

The molecular formula is $(C_3H_2Cl)_2$, or $C_6H_4Cl_2$.

3.67 Find the % composition of C and S from the analysis:

$$0.01665 \text{ g CO}_2 \times \frac{1 \text{ mol CO}_2}{44.01 \text{ g CO}_2} \times \frac{1 \text{ mol C}}{1 \text{ mol CO}_2} \times \frac{12.01 \text{ g C}}{1 \text{ mol C}} = 0.00454\underline{4} \text{ g C}$$

% C = (0.004544 g C ÷ 0.00796 g comp.) x 100% = 57.0̲9%

$$0.01196 \text{ g BaSO}_4 \times \frac{1 \text{ mol BaSO}_4}{233.39 \text{ g BaSO}_4} \times \frac{1 \text{ mol S}}{1 \text{ mol BaSO}_4} \times \frac{32.07 \text{ g S}}{1 \text{ mol S}}$$

$$= 0.00164\underline{3} \text{ g S}$$

% S = (0.001643 g S ÷ 0.00431 g comp.) x 100% = 38.1̲2%

% H = 100.00% - (57.09 + 38.12)% = 4.7̲9%

(continued)

We now obtain the empirical formula by calculating moles from the grams corresponding to each mass percentage of element:

$$57.09 \text{ g C} \times \frac{1 \text{ mol C}}{12.01 \text{ g C}} = 4.75\underline{4} \text{ mol C}$$

$$38.12 \text{ g S} \times \frac{1 \text{ mol S}}{32.07 \text{ g C}} = 1.18\underline{9} \text{ mol S}$$

$$4.79 \text{ g H} \times \frac{1 \text{ mol H}}{1.008 \text{ g H}} = 4.7\underline{52} \text{ mol H}$$

Dividing the moles of the elements by the smallest number (1.189), we obtain for C: 3.997, or 4; for S: 1.000, or 1; and for H: 3.996, or 4. Thus, the empirical formula is C_4H_4S (formula weight = 84). Because the formula weight was given as 84 amu, the molecular formula is also C_4H_4S.

3.69 For g $CaCO_3$, use this equation: $CaCO_3 + H_2C_2O_4 \rightarrow CaC_2O_4 + H_2O + CO_2$.

$$0.428 \text{ g CaC}_2\text{O}_4 \times \frac{1 \text{ mol CaC}_2\text{O}_4}{128.10 \text{ g CaC}_2\text{O}_4} \times \frac{1 \text{ mol CaCO}_3}{1 \text{ mol CaC}_2\text{O}_4} \times \frac{100.09 \text{ g CaCO}_3}{1 \text{ mol CaCO}_3}$$

$$= 0.33\underline{44} \text{ g CaCO}_3$$

$$\text{Mass\% CaCO}_3 = \frac{\text{mass CaCO}_3}{\text{mass limestone}} \times 100\% = \frac{0.3344 \text{ g}}{0.417 \text{ g}} \times 100\%$$

$$= 80.\underline{19} = 80.2\%$$

3.71 Calculate the theoretical yield using this equation: $2C_2H_4 + O_2 \rightarrow 2C_2H_4O$.

$$10.6 \text{ g C}_2\text{H}_4 \times \frac{1 \text{ mol C}_2\text{H}_4}{28.05 \text{ g C}_2\text{H}_4} \times \frac{2 \text{ mol C}_2\text{H}_4\text{O}}{2 \text{ mol C}_2\text{H}_4} \times \frac{44.05 \text{ g C}_2\text{H}_4\text{O}}{1 \text{ mol C}_2\text{H}_4\text{O}}$$

$$= 16.\underline{65} \text{ g C}_2\text{H}_4\text{O}$$

$$\text{\%yield} = \frac{\text{actual yield}}{\text{theoretical yield}} \times 100\% = \frac{9.91 \text{ g}}{16.65 \text{ g}} \times 100\% = 59.\underline{53} = 59.5\%$$

3.73   To find Zn, use these equations:

$$2C + O_2 \rightarrow 2CO \text{ and } ZnO + CO \rightarrow Zn + CO_2$$

Two mol C produces 2 mol CO; because 1 mol ZnO reacts with 1 mol CO, 2 mol ZnO will react with 2 mol CO. Thus, 2 mol C is equivalent to 2 mol ZnO, or 1 mol C is equivalent to 1 mol ZnO. Using this to calculate mass of C from mass of ZnO, we have

$$11.2 \text{ g ZnO} \times \frac{1 \text{ mol ZnO}}{81.39 \text{ g ZnO}} \times \frac{1 \text{ mol C}}{1 \text{ mol ZnO}} \times \frac{12.01 \text{ g C}}{1 \text{ mol C}} = 1.6\underline{5}2 \text{ g C}$$

Thus, all of the ZnO is used up in reacting with just 1.652 g of C, making ZnO the limiting reactant. Use the mass of ZnO to calculate the mass of Zn formed:

$$11.2 \text{ g ZnO} \times \frac{1 \text{ mol ZnO}}{81.39 \text{ g ZnO}} \times \frac{1 \text{ mol Zn}}{1 \text{ mol ZnO}} \times \frac{65.39 \text{ g Zn}}{1 \text{ mol Zn}} = 8.9\underline{9}8 = 9.00 \text{ g Zn}$$

■ Solutions to Cumulative-Skills Problems

3.75   Let y = the mass of CuO in the mixture. Then 0.500 g - y = the mass of Cu_2O in the mixture. Multiplying the appropriate conversion factors for Cu times the mass of each oxide will give one equation in one unknown for the mass of 0.425 g Cu:

$$0.425 = y \left[\frac{63.55 \text{ g Cu}}{79.55 \text{ g CuO}} \right] + (0.500 - y) \left[\frac{127.10 \text{ g Cu}}{143.10 \text{ g Cu}_2\text{O}} \right]$$

Simplifying the equation by dividing the conversion factors and combining terms gives:

$$0.425 = 0.798\underline{8}7 \text{ y} + 0.88819\underline{0} (0.500 - \text{y})$$

$$0.08932 \text{ y} = 0.01\underline{9}095$$

$$\text{y} = 0.2\underline{1}39 = 0.21 \text{ g} = \text{mass of CuO}$$

3.77   If one heme molecule contains one iron atom, then the number of moles of heme in 35.2 mg heme must be the same as the number of moles of iron in 3.19 mg of iron. Start by calculating the moles of Fe (= moles heme):

$$3.19 \times 10^{-3} \text{ g Fe} \times \frac{1 \text{ mol Fe}}{55.85 \text{ g Fe}} = 5.7\underline{1}2 \times 10^{-5} \text{ mol Fe or heme}$$

$$\text{Molar mass of heme} = \frac{35.2 \times 10^{-3} \text{ g}}{5.712 \times 10^{-5} \text{ mol}} = 61\underline{6}.2 = 616 \text{ g/mol}$$

The molecular weight of heme is 616 amu.

3.79   Use the data to find the molar mass of the metal and anion. Start with X_2.

Mass X_2 in MX = 4.52 g - 3.41 g = 1.11 g

Molar mass X_2 = 1.11 g ÷ 0.0158 mol = 70.$\underline{2}$5 g/mol

Molar mass X = 70.25 ÷ 2 = 35.$\underline{1}$4 = 35.1 g/mol

Thus X is Cl, chlorine.

Moles of M in 4.52 g MX = 0.0158 × 2 = 0.0316 mol

Molar mass of M = 3.41 g ÷ 0.0316 mol = 10$\underline{7}$.9 = 108 g/mol

Thus M is Ag, silver.

4. CHEMICAL REACTIONS

■ Answers to Review Questions

4.1   Some electrolyte solutions are strongly conducting because they are almost
       completely ionized, and others are weakly conducting because they are weakly
       ionized. The former solutions will have many more ions to conduct electricity than will
       the latter solutions if both are present at the same concentrations.

4.2   A strong electrolyte is an electrolyte that exists in solution almost entirely as ions. An
       example is NaCl. When NaCl dissolves in water, it dissolves almost completely to give
       Na^+ and Cl^- ions. A weak electrolyte is an electrolyte that dissolves in water to give a
       relatively small percentage of ions. An example is NH_3. When NH_3 dissolves in water,
       it reacts very little with the water, so the level of NH_3 is relatively high, and the level of
       NH_4^+ and OH^- ions is relatively low.

4.3   A spectator ion is an ion that does not take part in the reaction. In the following ionic
       reaction, the Na^+ and Cl^- are spectator ions:

$$Na^+(aq) + OH^-(aq) + H^+(aq) + Cl^-(aq) \rightarrow Na^+(aq) + Cl^-(aq) + H_2O(l)$$

4.4   A net ionic equation is an ionic equation from which spectator ions have been
       canceled. The value of such an equation is that it shows the reaction that actually
       occurs at the ionic level. An example is the ionic equation representing the reaction of
       calcium chloride ($CaCl_2$) with potassium carbonate (K_2CO_3).

$$CaCl_2(aq) + K_2CO_3(aq) \rightarrow CaCO_3(s) + 2\ KCl(aq):$$

$$Ca^{2+}(aq) + CO_3^{2-}(aq) \rightarrow CaCO_3(s) \qquad \text{(net ionic equation)}$$

4.5   The three major types of chemical reactions are precipitation reactions, acid-base reactions, and oxidation-reduction reactions. Oxidation-reduction reactions can be further classified as combination reactions, decomposition reactions, displacement reactions, and combustion reactions. Brief descriptions and examples of each are given below.

A precipitation reaction is a reaction that appears to involve the exchange of parts of the reactants. An example is: $2KCl\,(aq)\,+\,Pb(NO_3)_2(aq)\,\rightarrow\,2KNO_3(aq)\,+\,PbI_2(s)$.

An acid-base reaction, or neutralization reaction, results in an ionic compound and possibly water. An example is: $HCl(aq)\,+\,NaOH(aq)\,\rightarrow\,NaCl(aq)\,+\,H_2O(l)$.

A combination reaction is a reaction in which two substances combine to form a third substance. An example is: $2Na(s)\,+\,Cl_2(g)\,\rightarrow\,2NaCl(s)$.

A decomposition reaction is a reaction in which a single compound reacts to give two or more substances. An example is: $2HgO(s)\,\xrightarrow{\Delta}\,2Hg(l)\,+\,O_2(g)$.

A displacement reaction, or single replacement reaction, is a reaction in which an element reacts with a compound displacing an element from it. An example is: $Cu(s)\,+\,2AgNO_3(aq)\,\rightarrow\,2Ag(s)\,+\,Cu(NO_3)_2(aq)$.

A combustion reaction is a reaction of a substance with oxygen, usually with rapid release of heat to produce a flame. The products include one or more oxides. An example is: $CH_4(g)\,+\,2O_2(g)\,\rightarrow\,CO_2(g)\,+\,2H_2O(l)$.

4.6   To prepare crystalline AgCl and NaNO$_3$, first make solutions of AgNO$_3$ and NaCl by weighing equivalent molar amounts of both solid compounds. Then mix the two solutions together, forming a precipitate of silver chloride and a solution of soluble sodium nitrate. Filter off the silver chloride, and wash it with water to remove the sodium nitrate solution. Then allow it to dry to obtain pure crystalline silver chloride. Finally, take the filtrate containing the sodium nitrate, and evaporate it, leaving pure crystalline sodium nitrate.

4.7   An example of a neutralization reaction is

$$\underset{\text{acid}}{HBr}\;+\;\underset{\text{base}}{KOH}\;\rightarrow\;\underset{\text{salt}}{KBr}\;+\;H_2O(l)$$

4.8   An example of a polyprotic acid is carbonic acid, H_2CO_3. The successive neutralization is given by the following molecular equations:

$$H_2CO_3(aq)\,+\,NaOH(aq)\,\rightarrow\,NaHCO_3(aq)\,+\,H_2O(l)$$

$$NaHCO_3(aq)\,+\,NaOH(aq)\,\rightarrow\,Na_2CO_3(aq)\,+\,H_2O(l)$$

4.9   Since an oxidation-reduction reaction is an electron transfer reaction, one substance must lose the electrons and be oxidized while another substance must gain electrons and be reduced.

4.10   A displacement reaction is an oxidation-reduction reaction in which a free element reacts with a compound, displacing an element from it.

$$Cu(s) + 2AgNO_3(aq) \rightarrow 2Ag(s) + Cu(NO_3)_2(aq)$$

Ag^+ is the oxidizing agent, and Cu is the reducing agent.

■ Solutions to Practice Problems

Note on significant figures: If the final answer to a solution needs to be rounded off, it is given first with one nonsignificant figure, and the last significant figure is underlined. The final answer is then rounded to the correct number of significant figures. In multiple-step problems, intermediate answers are given with at least one nonsignificant figure; however, only the final answer has been rounded off.

4.15 a. Insoluble

   b. Soluble; The ions present would be NH_4^+ and SO_4^{2-}.

   c. Soluble; The ions present would be NH_4^+ and PO_4^{3-}.

   d. Soluble; The ions present would be Na^+ and CO_3^{2-}.

4.17 a. $H^+(aq) + OH^-(aq) \rightarrow H_2O(l)$

   b. $Ag^+(aq) + Br^-(aq) \rightarrow AgBr(s)$

   c. $S^{2-}(aq) + 2H^+(aq) \rightarrow H_2S(g)$

   d. $OH^-(aq) + NH_4^+(aq) \rightarrow NH_3(g) + H_2O(l)$

4.19 a. $FeSO_4(aq) + NaCl(aq) \rightarrow NR$

   b. $Na_2CO_3(aq) + CaCl_2(aq) \rightarrow CaCO_3(s) + 2NaCl(aq)$

   $$CO_3^{2-}(aq) + Ca^{2+}(aq) \rightarrow CaCO_3(s)$$

(continued)

c. $MgSO_4(aq) + 2NaOH(aq) \rightarrow Mg(OH)_2(s) + Na_2SO_4(aq)$

$\qquad Mg^{2+}(aq) + 2OH^-(aq) \rightarrow Mg(OH)_2(s)$

d. $NiCl_2(aq) + Li_2CO_3(aq) \rightarrow NiCO_3(s) + 2LiCl(aq)$

$\qquad Ni^{2+}(aq) + CO_3^{2-}(aq) \rightarrow NiCO_3(s)$

4.21 a. $Ba(NO_3)_2(aq) + LiSO_4(aq) \rightarrow BaSO_4(s) + 2LiNO_3(aq)$

$\qquad Ba^{2+}(aq) + SO_4^{2-}(aq) \rightarrow BaSO_4(s)$

b. $Ca(NO_3)_2(aq) + NaBr(aq) \rightarrow$ NR

c. $Al_2(SO_4)_3(aq) + 6NaOH(aq) \rightarrow 2Al(OH)_3(s) + 3Na_2SO_4(aq)$

$\qquad Al^{3+}(aq) + 3OH^-(aq) \rightarrow Al(OH)_3(s)$

d. $3CaBr_2(aq) + 2Na_3PO_4(aq) \rightarrow Ca_3(PO_4)_2(s) + 6NaBr(aq)$

$\qquad 3Ca^{2+}(aq) + 2PO_4^{3-}(aq) \rightarrow Ca_3(PO_4)_2(s)$

4.23 a. Weak acid   b. Weak base   c. Strong acid   d. Weak base

4.25 a. $NaOH(aq) + HClO_4(aq) \rightarrow H_2O(l) + NaClO_4(aq)$

$\qquad H^+(aq) + OH^-(aq) \rightarrow H_2O(l)$

b. $2HCl(aq) + Ba(OH)_2(aq) \rightarrow 2H_2O(l) + BaCl_2(aq)$

$\qquad H^+(aq) + OH^-(aq) \rightarrow H_2O(l)$

c. $2HC_2H_3O_2(aq) + Ca(OH)_2(aq) \rightarrow 2H_2O(l) + Ca(C_2H_3O_2)_2(aq)$

$\qquad HC_2H_3O_2(aq) + OH^-(aq) \rightarrow H_2O(l) + C_2H_3O_2^-(aq)$

d. $NH_3(aq) + HNO_2(aq) \rightarrow NH_4NO_2(aq)$

$\qquad NH_3(aq) + HNO_2(aq) \rightarrow NH_4^+(aq) + NO_2^-(aq)$

4.27 a. $2HBr(aq) + Ca(OH)_2(aq) \rightarrow 2H_2O(l) + CaBr_2(aq)$

   $H^+(aq) + OH^-(aq) \rightarrow H_2O(l)$

b. $3HNO_3(aq) + Al(OH)_3(s) \rightarrow 3H_2O(l) + Al(NO_3)_3(aq)$

   $3H^+(aq) + Al(OH)_3(s) \rightarrow 3H_2O(l) + Al^{3+}(aq)$

c. $2HCN(aq) + Ca(OH)_2(aq) \rightarrow 2H_2O(l) + Ca(CN)_2(aq)$

   $HCN(aq) + OH^-(aq) \rightarrow H_2O(l) + CN^-(aq)$

d. $HCN(aq) + LiOH(aq) \rightarrow H_2O(l) + LiCN(aq)$

   $HCN(aq) + OH^-(aq) \rightarrow H_2O(l) + CN^-(aq)$

4.29 a. Molecular equation: $FeS(aq) + 2HBr(aq) \rightarrow FeBr_2(aq) + H_2S(g)$

   Ionic equation: $S^{2-}(aq) + 2H^+(aq) \rightarrow H_2S(g)$

b. Molecular equation: $MgCO_3(s) + 2HNO_3(aq) \rightarrow$
   $$Mg(NO_3)_2(aq) + CO_2(g) + H_2O(l)$$

   Ionic equation: $MgCO_3(s) + 2H^+(aq) \rightarrow Mg^{2+}(aq) + CO_2(g) + H_2O(l)$

c. Molecular equation: $K_2SO_3(aq) + H_2SO_4(aq) \rightarrow K_2SO_4(aq) + SO_2(g) + H_2O(l)$

   Ionic equation: $SO_3^{2-}(aq) + 2H^+(aq) \rightarrow SO_2(g) + H_2O(l)$

4.31 a. Because the phosphate ion (PO_4^{3-}) has a net oxidation number of -3, the oxidation number of Ga = +3.

b. Because all five O's = a total of -10, the oxidation number of Nb = +5.

c. Because the 3 O's = a total of -6 and K = +1, the oxidation number of Br = +5.

d. Because the 4 O's = a total of -8 and the 2 K's = +2, the oxidation number of Mn = +6.

4.33 a. Because the charge of -1 = [x_N + 2 (from 2 H's)], x_N must equal -3.

b. Because the charge of -1 = [x_I - 4 (from 2 O's)], x_I must equal +3.

c. Because the charge of -1 = [x_{Al} - 8 (4 O's) + 4 (4 H's)], x_{Al} must equal +3.

d. Because the charge of -2 = [x_P - 8 (4 O's) + 1 (1 H)], x_P must equal +5.

4.35 a. Phosphorus changes from an oxidation number of zero in P_4 to +5 in P_4O_{10}, losing electrons and acting as a reducing agent. Oxygen changes from an oxidation number of zero in O_2 to -2 in P_4O_{10}, gaining electrons and acting as an oxidizing agent.

b. Cobalt changes from an oxidation number of zero in Co(s) to +2 in $CoCl_2$, losing electrons and acting as a reducing agent. Chlorine changes from an oxidation number of zero in Cl_2 to -1 in $CoCl_2$, gaining electrons and acting as an oxidizing agent.

4.37 a. First identify the species being oxidized and reduced, and assign the appropriate oxidation states. Since $CuCl_2$ and $AlCl_3$ are both soluble ionic compounds, Cl^- is a spectator ion and can be removed from the equation. The resulting net ionic equation is

$$\overset{+2}{Cu^{2+}}(aq) + \overset{0}{Al}(s) \rightarrow \overset{+3}{Al^{3+}}(aq) + \overset{0}{Cu}(s)$$

Next write the half-reactions in an unbalanced form.

$$Al \rightarrow Al^{3+} \qquad \text{(oxidation)}$$
$$Cu^{2+} \rightarrow Cu \qquad \text{(reduction)}$$

Then balance the charge in each equation by adding electrons to the more positive side to create balanced half-reactions.

$$Al \rightarrow Al^{3+} + 3e^- \qquad \text{(oxidation half-reaction)}$$
$$Cu^{2+} + 2e^- \rightarrow Cu \qquad \text{(reduction half-reaction)}$$

Multiply each half-reaction by a factor that will cancel out the electrons.

$$2 \times (Al \rightarrow Al^{3+} + 3e^-)$$
$$\underline{3 \times (Cu^{2+} + 2e^- \rightarrow Cu)}$$
$$3Cu^{2+} + 2Al + 6e^- \rightarrow 2Al^{3+} + 3Cu + 6e^-$$

Therefore, the balanced oxidation-reduction reaction is

$$3Cu^{2+} + 2Al \rightarrow 2Al^{3+} + 3Cu$$

Finally, add six Cl^- ions to each side, and add phase labels. The resulting balanced equation is

$$3CuCl_2(aq) + 2Al(s) \rightarrow 2AlCl_3(aq) + 3Cu(s)$$

(continued)

b. First identify the species being oxidized and reduced and assign the appropriate oxidation states.

$$\overset{+3}{Cr^{3+}}(aq) + \overset{0}{Zn}(s) \rightarrow \overset{0}{Cr}(s) + \overset{+2}{Zn^{2+}}(aq)$$

Next write the half-reactions in an unbalanced form.

$Zn \rightarrow Zn^{2+}$ (oxidation)

$Cr^{3+} \rightarrow Cr$ (reduction)

Then balance the charge in each equation by adding electrons to the more positive side to create balanced half-reactions.

$Zn \rightarrow Zn^{2+} + 2e^-$ (oxidation half-reaction)

$Cr^{3+} + 3e^- \rightarrow Cr$ (reduction half-reaction)

Multiply each half-reaction by a factor that will cancel out the electrons.

$$3 \times (Zn \rightarrow Zn^{2+} + 2e^-)$$
$$\underline{2 \times (Cr^{3+} + 3e^- \rightarrow Cr)}$$
$$2Cr^{3+} + 3Zn + \cancel{6e^-} \rightarrow 2Cr + 3Zn^{2+} + \cancel{6e^-}$$

Therefore, the balanced oxidation-reduction reaction, including phase labels, is

$$2Cr^{3+}(aq) + 3Zn(s) \rightarrow 2Cr(s) + 3Zn^{2+}(aq)$$

4.39   Molarity $= \dfrac{\text{moles solute}}{\text{liters solution}} = \dfrac{0.0147 \text{ mol}}{0.0400 \text{ L}} = 0.36\underline{7}5 = 0.368 \ M$

4.41 $0.150 \text{ mol CuSO}_4 \times \dfrac{1 \text{ L solution}}{0.185 \text{ mol CuSO}_4} = 0.81\underline{0}8 \text{ L (811 mL soln)}$

4.43   From the molarity, 1 L of heme solution is equivalent to 0.0019 mol of heme solute. Before starting the calculation, note that 75 mL of solution is equivalent to 75×10^{-3} L of solution:

$$75 \times 10^{-3} \text{ L soln} \times \dfrac{0.0019 \text{ mol heme}}{1 \text{ L solution}} = 1.\underline{4}25 \times 10^{-4} = 1.4 \times 10^{-4} \text{ mol heme}$$

4.45   Multiply the volume of solution by molarity to convert it to moles; then convert to mass of solute by multiplying by the molar mass:

$$50 \times 10^{-3} \text{ L soln} \times \frac{0.025 \text{ mol Na}_2\text{Cr}_2\text{O}_7}{1 \text{ L solution}} \times \frac{262.0 \text{ g Na}_2\text{Cr}_2\text{O}_7}{1 \text{ mol Na}_2\text{Cr}_2\text{O}_7}$$

$$= 0.3\underline{2}75 = 0.33 \ g \text{ Na}_2\text{Cr}_2\text{O}_7$$

4.47   Use the rearranged version of the dilution formula to calculate the initial volume of 15.8 M HNO$_3$ required:

$$V_i = \frac{M_f V_f}{M_i} = \frac{0.16 \ M \times 1000 \text{ mL}}{15.8 \ M} = 1\underline{0}.12 = 10. \text{ mL}$$

4.49   Use the appropriate conversion factors to convert the mass of BaSO$_4$ to the mass of Ba^{2+} ions:

$$0.513 \text{ g BaSO}_4 \times \frac{1 \text{ mol BaSO}_4}{233.40 \text{ g BaSO}_4} \times \frac{1 \text{ mol Ba}^{2+}}{1 \text{ mol BaSO}_4} \times \frac{137.33 \text{ g Ba}^{2+}}{1 \text{ mol Ba}^{2+}}$$

$$= 0.30\underline{1}84 \text{ g Ba}^{2+}$$

Then calculate the percentage of barium in the 458 mg (0.458 g) compound:

$$\frac{0.30184 \text{ g Ba}^{2+}}{0.458 \text{ g}} \times 100\% = 65.\underline{9}039 = 65.9\% \text{ Ba}^{2+}$$

4.51   First calculate the moles of chlorine in the compound:

$$0.3048 \text{ g AgCl} \times \frac{1 \text{ mol AgCl}}{143.32 \text{ g AgCl}} \times \frac{1 \text{ mol Cl}^-}{1 \text{ mol AgCl}} = 0.002126\underline{7} \text{ mol Cl}^-$$

Then calculate the g Fe^{x+} from the g Cl$^-$:

$$\text{g Fe}^{x+} = 0.1348 \text{ g comp} - \left(0.0021267 \text{ mol Cl}^- \times \frac{35.45 \text{ g Cl}^-}{1 \text{ mol Cl}^-} \right)$$

$$= 0.0594\underline{0}8 \text{ g Fe}^{x+}$$

(continued)

Now calculate the moles of Fe^{x+} using the molar mass:

$$0.059408 \text{ g } Fe^{x+} \times \frac{1 \text{ mol } Fe^{x+}}{55.85 \text{ g } Fe^{x+}} = 0.0010637 \text{ mol } Fe^{x+}$$

Finally, divide the mole numbers by the smallest mole number:

$$\text{For Cl:} \quad \frac{0.002127 \text{ mol } Cl^-}{0.0010637 \text{ mol}} = 2.00; \quad \text{for } Fe^{x+}: \quad \frac{0.0010637 \text{ mol } Fe^{x+}}{0.0010637 \text{ mol}} = 1.00$$

Thus the formula is $FeCl_2$.

4.53 Using molarity, convert the volume of Na_2CO_3 to moles of Na_2CO_3; then use the equation to convert to moles of HNO_3 and finally to volume:

$$2HNO_3 + Na_2CO_3 \rightarrow 2NaNO_3 + H_2O + CO_2$$

$$42.4 \times 10^{-3} \text{ L } Na_2CO_3 \times \frac{0.150 \text{ mol } Na_2CO_3}{1 \text{ L soln}} \times \frac{2 \text{ mol } HNO_3}{1 \text{ mol } Na_2CO_3}$$

$$\times \frac{1 \text{ L } HNO_3}{0.250 \text{ mol } HNO_3} = 0.05088 \text{ L } (50.9 \text{ mL}) \text{ } HNO_3$$

4.55 First find the mass of H_2O_2 required to react with $KMnO_4$.

$$5H_2O_2 + 2KMnO_4 + 3H_2SO_4 \rightarrow 5O_2 + 2MnSO_4 + K_2SO_4 + 8H_2O$$

$$51.7 \times 10^{-3} \text{ L } KMnO_4 \times \frac{0.145 \text{ mol } KMnO_4}{1 \text{ L soln}} \times \frac{5 \text{ mol } H_2O_2}{2 \text{ mol } KMnO_4} \times \frac{34.02 \text{ g } H_2O_2}{1 \text{ mol } H_2O_2}$$

$$= 0.6375 \text{ g } H_2O_2$$

$$\% \text{ } H_2O_2 = (\text{mass } H_2O_2 \div \text{ mass sample}) \times 100\%$$

$$= (0.6375 \text{ g} \div 20.0 \text{ g}) \times 100\% = 3.187 = 3.19\%$$

■ Solutions to General Problems

4.57 For the reaction of magnesium metal and hydrobromic acid, the equations are as follows.

Molecular equation: $Mg(s) + 2HBr(aq) \rightarrow H_2(g) + MgBr_2(aq)$

Ionic equation: $Mg(s) + 2H^+(aq) \rightarrow H_2(g) + Mg^{2+}(aq)$

4.59 For the reaction of nickel(II) sulfate and lithium hydroxide, the equations are as follows.

Molecular equation: $NiSO_4(aq) + 2LiOH(aq) \rightarrow Ni(OH)_2(s) + Li_2SO_4(aq)$

Ionic equation: $Ni^{2+}(aq) + 2OH^-(aq) \rightarrow Ni(OH)_2(s)$

4.61 a. Molecular equation: $Sr(OH)_2(aq) + 2HC_2H_3O_2(aq) \rightarrow$
$$Sr(C_2H_3O_2)_2(aq) + 2H_2O(l)$$

Ionic equation: $HC_2H_3O_2(aq) + OH^-(aq) \rightarrow C_2H_3O_2^-(aq) + H_2O(l)$

b. Molecular equation: $NH_4I(aq) + NaOH(aq) \rightarrow NaI(aq) + NH_3(g) + H_2O(l)$

Ionic equation: $NH_4^+(aq) + OH^-(aq) \rightarrow NH_3(g) + H_2O(l)$

c. Molecular equation: $NaNO_3(aq) + CsCl(aq) \rightarrow NR$
$$\text{(NaCl and CsNO}_3 \text{ are soluble.)}$$

d. Molecular equation: $NH_4I(aq) + AgNO_3(aq) \rightarrow NH_4NO_3(aq) + AgI(s)$

Ionic equation: $I^-(aq) + Ag^+(aq) \rightarrow AgI(s)$

4.63 For each preparation, the compound to be prepared is given first, followed by the compound from which it is to be prepared. Then the method of preparation is given, followed by the molecular equation for the preparation reaction. Steps such as evaporation, etc., are not given in the molecular equation.

a. To prepare $CuCl_2$ from $CuSO_4$, add a solution of $BaCl_2$ to a solution of the $CuSO_4$, precipitating $BaSO_4$. The $BaSO_4$ can be filtered off, leaving aqueous $CuCl_2$, which can be obtained in solid form by evaporation of the solution. Molecular equation:

$$CuSO_4(aq) + BaCl_2(aq) \rightarrow BaSO_4(s) + CuCl_2(aq)$$

(continued)

b. To prepare $Ca(C_2H_3O_2)_2$ from $CaCO_3$, add a solution of acetic acid, $HC_2H_3O_2$, to the solid $CaCO_3$, forming CO_2, H_2O, and aqueous $Ca(C_2H_3O_2)_2$. The aqueous $Ca(C_2H_3O_2)_2$ can be converted to the solid form by evaporation of the solution, which also removes the CO_2 and H_2O products. Molecular equation:

$$CaCO_3(s) + 2HC_2H_3O_2(aq) \rightarrow Ca(C_2H_3O_2)_2(aq) + CO_2(g) + H_2O(l)$$

c. To prepare $NaNO_3$ from Na_2SO_3, add a solution of nitric acid, HNO_3, to the solid Na_2SO_3, forming SO_2, H_2O, and aqueous $NaNO_3$. The aqueous $NaNO_3$ can be converted to the solid by evaporation of the solution, which also removes the SO_2 and H_2O products. Molecular equation:

$$Na_2SO_3(s) + 2HNO_3(aq) \rightarrow 2NaNO_3(aq) + SO_2(g) + H_2O(l)$$

d. To prepare $MgCl_2$ from $Mg(OH)_2$, add a solution of hydrochloric acid (HCl) to the solid $Mg(OH)_2$, forming H_2O and aqueous $MgCl_2$. The aqueous $MgCl_2$ can be converted to the solid form by evaporation of the solution. Molecular equation:

$$Mg(OH)_2(s) + 2HCl(aq) \rightarrow MgCl_2(aq) + 2H_2O(l)$$

4.65   a. Decomposition   b. Decomposition   c. Combination   d. Displacement

4.67   Divide the mass of $CaCl_2$ by its molar mass and volume to find molarity:

$$4.50 \text{ g CaCl}_2 \times \frac{1 \text{ mol CaCl}_2}{110.98 \text{ g CaCl}_2} \times \frac{1}{1.000 \text{ L soln}} = 0.04054 = 0.0405 \text{ M CaCl}_2$$

The $CaCl_2$ dissolves to form Ca^{2+} and $2Cl^-$ ions. Therefore, the molarities of the ions are 0.0405 M Ca^{2+} and 2 x 0.04054, or 0.0811, M Cl^- ions.

4.69   Divide the mass of $K_2Cr_2O_7$ by its molar mass and volume to find molarity. Then calculate the volume needed to prepare 1.00L of a 0.100 M solution.

$$115 \text{ g K}_2Cr_2O_7 \times \frac{1 \text{ mol K}_2Cr_2O_7}{294.20 \text{ g K}_2Cr_2O_7} = 0.39089 \text{ M}$$

$$\text{Molarity} = \frac{0.39089 \text{ mol K}_2Cr_2O_7}{1.00 \text{ L}} = 0.39089 \text{ M}$$

$$V_i = \frac{M_f V_f}{M_i} = \frac{0.100 \text{ M} \times 1.00 \text{ L}}{0.39089 \text{ M}} = 0.2558 \text{ L (256 mL)}$$

4.71 Assume a volume of 1.000 L (1000 cm³) for the 6.00% NaBr solution, and convert to moles and then to molarity.

$$1000 \text{ cm}^3 \times \frac{1.046 \text{ g soln}}{1 \text{ cm}^3} \times \frac{6.00 \text{ g NaBr}}{100 \text{ g soln}} \times \frac{1 \text{ mol NaBr}}{102.89 \text{ g NaBr}} = 0.60\underline{9}9 \text{ mol}$$

$$\text{Molarity NaBr} = \frac{0.6099 \text{ mol NaBr}}{1.000 \text{ L}} = 0.60\underline{9}9 = 0.610 \text{ } M$$

4.73 First calculate the moles of $BaCl_2$:

$$1.128 \text{ g BaSO}_4 \times \frac{1 \text{ mol BaSO}_4}{233.40 \text{ g BaSO}_4} \times \frac{1 \text{ mol BaCl}_2}{1 \text{ mol BaSO}_4} = 0.004832\underline{9} \text{ mol BaCl}_2$$

Then calculate the molarity from the moles and volume (0.0500 L):

$$\text{Molarity} = \frac{0.0048329 \text{ mol BaCl}_2}{0.0500 \text{ L}} = 0.096\underline{6}58 = 0.0967 \text{ } M$$

4.75 First calculate the g of thallium(I) sulfate:

$$0.2122 \text{ g TlI} \times \frac{1 \text{ mol TlI}}{331.28 \text{ g TlI}} \times \frac{1 \text{ mol Tl}_2\text{SO}_4}{2 \text{ mol TlI}} \times \frac{504.83 \text{ g Tl}_2\text{SO}_4}{1 \text{ mol Tl}_2\text{SO}_4}$$

$$= 0.161\underline{6}8 \text{ g Tl}_2\text{SO}_4$$

Then calculate the % Tl_2SO_4 in the rat poison:

$$\%\text{Ti}_2\text{SO}_4 = \frac{0.16168 \text{ g}}{0.7590 \text{ g}} \times 100\% = 21.3\underline{0}1 = 21.30\%$$

4.77 From the equations $NH_3 + HCl \rightarrow NH_4Cl$ and $NaOH + HCl \rightarrow NaCl + H_2O$, we write

Mol NH_3 = mol $HCl(NH_3)$

Mol NaOH = mol HCl(NaOH)

(continued)

We can calculate the mol NaOH and the sum [mol HCl(NH₃) + mol HCl(NaOH)] from the titration data. Because the sum = mol NH₃ + mol NaOH, we can calculate the unknown mol of NH₃ from the difference: Mol NH₃ = sum - mol NaOH.

$$\text{Mol HCl (NaOH)} + \text{mol HCl (NH}_3) = 0.0463 \text{ L} \times \frac{0.0213 \text{ mol HCl}}{1.000 \text{ L}}$$

$$= 0.009862 \text{ mol HCl}$$

$$\text{Mol NaOH} = 0.0443 \text{ L} \times \frac{0.128 \text{ mol NaOH}}{1.000 \text{ L}} = 0.005670 \text{ mol NaOH}$$

Mol HCl(NH₃) = 0.009862 mol - 0.005670 mol = 0.004192 mol

Mol NH₃ = mol HCl(NH₃) = 0.004192 mol NH₃

Because all the N in the (NH₄)₂SO₄ was liberated as and titrated as NH₃, the amount of N in the fertilizer is equal to the amount of N in the NH₃. Thus the moles of NH₃ can be used to calculate the mass percentage of N in the fertilizer:

$$0.004192 \text{ mol NH}_3 \times \frac{1 \text{ mol N}}{1 \text{ mol NH}_3} \times \frac{14.01 \text{ g N}}{1 \text{ mol N}} = 0.05873 \text{ g N}$$

$$\%N = \frac{\text{mass N}}{\text{mass fert.}} \times 100\% = \frac{0.05873 \text{ g N}}{0.608 \text{ g}} \times 100\% = 9.659 = 9.66\%$$

■ Solutions to Cumulative-Skills Problems

4.79 For this reaction, the formulas are listed first, followed by the molecular and net ionic equations, the names of the products, and the molecular equation for another reaction giving the same precipitate.

Lead(II) nitrate is Pb(NO₃)₂, and cesium sulfate is Cs₂SO₄.

Molecular equation: Pb(NO₃)₂(aq) + Cs₂SO₄(aq) → PbSO₄(s) + 2CsNO₃(aq)

Net ionic equation: Pb²⁺(aq) + SO₄²⁻(aq) → PbSO₄(s)

PbSO₄ is lead(II) sulfate, and CsNO₃ is cesium nitrate.

Molecular equation: Pb(NO₃)₂(aq) + Na₂SO₄(aq) → PbSO₄(s) + 2NaNO₃(aq)

4.81   Use the density, formula weight, and percentage to convert to molarity. Then combine the 0.200 mol with mol/L to obtain the volume in liters.

$$\frac{0.807 \text{ g soln}}{1 \text{ mL}} \times \frac{0.940 \text{ g ethanol}}{1.00 \text{ g soln}} \times \frac{1 \text{ mol ethanol}}{46.07 \text{ g ethanol}} \times \frac{1000 \text{ mL}}{1 \text{ L}}$$

$$= \frac{16.465 \text{ mol ethanol}}{1 \text{ L ethanol}}$$

$$\text{L ethanol} = 0.200 \text{ mol ethanol} \times \frac{1 \text{ L ethanol}}{16.465 \text{ mol ethanol}} = 0.012146 = 0.0121 \text{ L}$$

4.83   Convert the 2.183 g of Ag to mol AgI, which is chemically equivalent to moles of KI. Use that to calculate the molarity of the KI.

$$2.183 \text{ g AgI} \times \frac{1 \text{ mol AgI}}{234.77 \text{ g AgI}} = 9.2984 \times 10^{-3} \text{ mol AgI (eq to mol KI)}$$

$$\text{Molarity} = \frac{9.2984 \times 10^{-3} \text{ mol KI}}{0.0100 \text{ L}} = 0.9298 = 0.930 \text{ M}$$

4.85   Convert the 6.026 g of $BaSO_4$ to mol $BaSO_4$; then from the equation deduce that 3 mol $BaSO_4$ is equivalent to 1 mol $M_2(SO_4)_3$ and is equivalent to 2 mol of M. Use that with 1.200 g of the metal M to calculate the atomic weight of M.

$$6.026 \text{ g BaSO}_4 \times \frac{1 \text{ mol BaSO}_4}{233.40 \text{ g BaSO}_4} \times \frac{2 \text{ mol M}}{3 \text{ mol BaSO}_4} = 0.017212 \text{ mol } M$$

$$\text{Atomic wt of M in g/mol} = \frac{1.200 \text{ g M}}{0.017212 \text{ mol M}} = 69.719 \text{ g/mol (gallium)}$$

4.87   Use the density, formula weight, percentage, and volume to convert to mol H_3PO_4. Then, from the equation $P_4O_{10} + 6H_2O \rightarrow 4H_3PO_4$, deduce that 4 mol H_3PO_4 is equivalent to 1 mol of P_4O_{10}, and use that to convert to mol P_4O_{10}.

$$1500 \text{ mL} \times \frac{1.025 \text{ g soln}}{1 \text{ mL}} \times \frac{0.0500 \text{ g H}_3\text{PO}_4}{1 \text{ g soln}} \times \frac{1 \text{ mol H}_3\text{PO}_4}{98.00 \text{ g H}_3\text{PO}_4}$$

$$= 0.7844 \text{ mol H}_3\text{PO}_4$$

(continued)

$$0.7844 \text{ mol } H_3PO_4 \times \frac{1 \text{ mol } P_4O_{10}}{4 \text{ mol } H_3PO_4} = 0.196\underline{1} \text{ mol } P_4O_{10}$$

$$\text{Mass } P_4O_{10} = 0.1961 \text{ mol } P_4O_{10} \times \frac{283.92 \text{ g } P_4O_{10}}{1 \text{ mol } P_4O_{10}}$$

$$= 55.\underline{6}77 = 55.7 \text{ g } P_4O_{10}$$

4.89 Use the formula weight of $Al_2(SO_4)_3$ to convert to mol $Al_2(SO_4)_3$. Then deduce from the equation that 1 mol $Al_2(SO_4)_3$ is equivalent to 3 mol H_2SO_4, and calculate the moles of H_2SO_4 needed. Combine density, percentage, and formula weight to obtain molarity of H_2SO_4. Then combine molarity and moles to obtain volume.

$$37.4 \text{ g } Al_2(SO_4)_3 \times \frac{1 \text{ mol } Al_2(SO_4)_3}{342.17 \text{ g } Al_2(SO_4)_3} \times \frac{3 \text{ mol } H_2SO_4}{1 \text{ mol } Al_2(SO_4)_3}$$

$$= 0.327\underline{9} \text{ mol } H_2SO_4$$

$$\frac{1.104 \text{ g soln}}{1 \text{ mL}} \times \frac{15.0 \text{ g } H_2SO_4}{100 \text{ g soln}} \times \frac{1 \text{ mol } H_2SO_4}{98.09 \text{ g } H_2SO_4} \times \frac{1000 \text{ mL}}{1 \text{ L}}$$

$$= 1.6\underline{8}8 \text{ mol } H_2SO_4 / L$$

$$0.3279 \text{ mol } H_2SO_4 \times \frac{1 \text{ L } H_2SO_4}{1.688 \text{ mol } H_2SO_4} = 0.194\underline{2} = 0.194 \text{ L (194 mL)}$$

4.91 The equations for the neutralization are $2HCl + Mg(OH)_2 \rightarrow MgCl_2 + 2H_2O$ and $3HCl + Al(OH)_3 \rightarrow AlCl_3 + 3H_2O$. Calculate the moles of HCl, and set equal to the total moles of hydroxide ion, OH⁻.

$$0.0485 \text{ L HCl} \times \frac{0.187 \text{ mol HCl}}{1 \text{ L HCl}} = 0.00906\underline{9}5 \text{ mol HCl}$$

$$0.0090695 \text{ mol HCl} = 2 \text{ [mol } Mg(OH)_2] + 3 \text{ [mol } Al(OH)_3]$$

Rearrange the last equation, and solve for the moles of $Al(OH)_3$.

(Eq 1) mol $Al(OH)_3 = 0.0030231 \text{ mol HCl} - 2/3 \text{ [mol } Mg(OH)_2]$

The total mass of chloride salts is equal to the sum of the masses of $MgCl_2$ (molar mass = 95.21 g/mol) and $AlCl_3$ (molar mass = 133.33 g/mol).

$$[95.21 \text{ g/mol} \times \text{mol } MgCl_2] + [133.33 \text{ g/mol} \times \text{mol } AlCl_3] = 0.4200 \text{ g}$$

(continued)

Since the mol $Mg(OH)_2$ = mol $MgCl_2$, and mol $Al(OH)_3$ = mol $AlCl_3$, you can substitute these quantities into the last equation and get

$$[95.21 \text{ g/mol} \times \text{mol } Mg(OH)_2] + [133.33 \text{ g/mol} \times \text{mol } Al(OH)_3] = 0.4200 \text{ g}$$

Substitute Eq 1 into this equation for the mol of $Al(OH)_3$:

$$[95.21 \text{ g/mol} \times \text{mol } Mg(OH)_2]$$
$$+ [133.33 \text{ g/mol} \times (0.0030231 \text{ mol} - 2/3 \text{ mol } Mg(OH)_2)] = 0.4200 \text{ g}$$

Rearrange the equation, and solve for the moles of $Mg(OH)_2$.

$$6.323 \text{ mol } Mg(OH)_2 + 0.403070 = 0.4200$$

$$\text{mol } Mg(OH)_2 = 0.01693 \div 6.323 = 0.0026728 \text{ mol}$$

Calculate the mass of $Mg(OH)_2$ in the antacid tablet (molar mass = 58.33 g/mol).

$$0.0026728 \text{ mol } Mg(OH)_2 \times \frac{58.33 \text{ g } Mg(OH)_2}{1 \text{ mol } Mg(OH)_2} = 0.15590 \text{ g } Mg(OH)_2$$

Use Eq 1 to find the moles and mass of $Al(OH)_3$ (molar mass 78.00 g/mol) in a similar fashion.

$$\text{Mol } Al(OH)_3 = 0.0030231 - 2/3[0.0026728 \text{ mol } Mg(OH)_2] = 0.0012412 \text{ mol}$$

$$0.0012412 \text{ mol } Al(OH)_3 \times \frac{78.00 \text{ g } Al(OH)_3}{1 \text{ mol } Al(OH)_3} = 0.096681 \text{ g } Al(OH)_3$$

The percent $Mg(OH)_2$ in the antacid is

$$\% \ Mg(OH)_2 = [0.15590 \text{ g} \div (0.15590 + 0.096681) \text{ g}] \times 100\%$$

$$= 61.72 = 61.7\%$$

5. THE GASEOUS STATE

■ Answers to Review Questions

5.1   Pressure is the force exerted per unit area of surface. Force is further defined as mass
       multiplied by acceleration. The SI unit of mass is kg, of acceleration is m/s^2, and of
       area is m^2. Therefore, the SI unit of pressure (Pascal) is given by

$$\text{Pressure} = \frac{\text{force}}{\text{area}} = \frac{\text{mass x acceleration}}{\text{area}} = \frac{\text{kg x m/s}^2}{\text{m}^2} = \frac{\text{kg}}{\text{m} \cdot \text{s}^2} = \text{Pa}$$

5.2   The general relationship between the pressure (P) and the height (h) of the liquid in a
       manometer is $P = dgh$. Therefore, the variables that determine the height of the liquid
       in a manometer are the density (d) of the liquid and the pressure of the gas being
       measured. The acceleration of gravity (g) is a constant, 9.81 m/s^2.

5.3   First find the equivalent of absolute zero on the Fahrenheit scale. Converting
       -273.15°C to degrees Fahrenheit, you obtain -459.67°F. Since the volume of a gas
       varies linearly with the temperature, you get the following linear relationship.

$$V = a + bt_F$$

where t_F is the temperature on the Fahrenheit scale. Since the volume of a gas is zero
at absolute zero, you get

$$0 = a + b(-459.67), \text{ or } a = 459.67b$$

(continued)

The equation can now be rewritten as

$$V = 459.67b + bt = b(459.67 + t_F) = bT_F$$

where T_F is the temperature in the new absolute scale based on the Fahrenheit scale. The relationship is

$$T_F = t_F + 459.67$$

5.4   Avogadro's law states that equal volumes of any two gases at the same temperature and pressure contain the same number of molecules. The law of combining volumes states that the volumes of reactant gases at a given pressure and temperature are in ratios of small whole numbers. The combining-volume law may be explained from Avogadro's law using the reaction $N_2 + 3H_2 \rightarrow 2NH_3$ as follows: In Avogadro's terms, this equation says that Avogadro's number of N_2 molecules reacts with three times Avogadro's number of H_2 molecules to form two times Avogadro's number of NH_3 molecules. From Avogadro's law, it follows that one volume of N_2 reacts with three volumes of H_2 to form two volumes of NH_3. This result is true for all gas reactions.

5.5   The molar gas volume, V_m, is the volume of one mole of gas at any given temperature and pressure. At standard conditions (STP), the molar gas volume equals 22.4 L.

5.6   Boyle's law ($V \propto 1/P$) and Charles's law ($V \propto T$) can be combined and expressed in a single statement: The volume occupied by a given amount of gas is proportional to the absolute temperature divided by the pressure. In equation form, this is

$$V = \text{constant} \times \frac{T}{P}$$

The constant is independent of temperature and pressure but does depend on the amount of gas. For one mole, the constant will have a specific value denoted as R. The molar volume, V_m, is

$$V_m = R \times \frac{T}{P}$$

Because V_m has the same value for all gases, we can write this equation for n moles of gas if we multiply both sides by n. This yields the equation

$$nV_m = \frac{nRT}{P}$$

Because V_m is the volume per mole, nV_m is the total volume V. Substituting gives

$$V = \frac{nRT}{P} \quad \text{or} \quad PV = nRT$$

5.7   Use the value of R from Table 5.5 and the conversion factor 1 atm = 760 mmHg. This gives

$$0.082058 \frac{L \cdot atm}{K \cdot mol} \times \frac{760 \text{ mmHg}}{1 \text{ atm}} = 62.36\underline{4}0 = 62.364 \frac{L \cdot mmHg}{K \cdot mol}$$

5.8   The postulates and supporting evidence are the following:

(1) Gases are composed of molecules whose sizes are negligible compared with the distance between them.

(2) Molecules move randomly in straight lines in all directions and at various speeds.

(3) The forces of attraction or repulsion between two molecules (intermolecular forces) in a gas are very weak or negligible, except when they collide.

(4) When molecules collide with one another, the collisions are elastic.

(5) The average kinetic energy of a molecule is proportional to the absolute temperature.

One example of evidence that supports the kinetic theory of gases is Boyle's law. Constant temperature means the average molecular force from collision remains constant. If you increase the volume, you decrease the number of collisions per unit wall area, thus lowering the pressure in accordance with Boyle's law. Another example is Charles's law. If you raise the temperature, you increase the average molecular force of a collision with the wall, thus increasing the pressure. For the pressure to remain constant, it is necessary for the volume to increase so the frequency of collisions with the wall decreases. Thus, when you raise the temperature of a gas while keeping the pressure constant, the volume increases in accordance with Charles's law.

5.9   According to kinetic theory, the pressure of a gas results from the bombardment of container walls by molecules.

5.10   The rms speed of a molecule equals $(3RT/M_m)^{1/2}$, where M_m is the molar mass of the gas. The rms speed does not depend on the molar volume.

5.11   A gas appears to diffuse more slowly because it never travels very far in one direction before it collides with another molecule and moves in another direction. Thus it must travel a very long, crooked path as the result of collisions.

5.12   Effusion is the process in which a gas flows through a small hole in a container. It results from the gas molecules encountering the hole by chance rather than by colliding with the walls of the container. The faster the molecules move, the more likely they are to encounter the hole. Thus the rate of effusion depends on the average molecular speed, which depends inversely on molecular mass.

5.13   The behavior of a gas begins to deviate significantly from that predicted by the ideal gas law at high pressures and relatively low temperatures.

5.14   The constant "*a*" is the proportionality constant in the van der Waals equation related to intermolecular forces The term "*nb*" represents the volume occupied by *n* moles of molecules.

■ Solutions to Practice Problems

Note on significant figures: If the final answer to a solution needs to be rounded off, it is given first with one nonsignificant figure, and the last significant figure is underlined. The final answer is then rounded to the correct number of significant figures. In multiple-step problems, intermediate answers are given with at least one nonsignificant figure; however, only the final answer has been rounded off.

5.19   Use the conversion factor 1 atm = 760 mmHg.

$$0.053 \text{ atm } \times \frac{760 \text{ mmHg}}{1 \text{ atm}} = 40.2 = 40. \text{ mmHg}$$

5.21   Using Boyle's law, solve for V_f of the neon gas at 1.316 atm pressure.

$$V_f = V_i \times \frac{P_i}{P_f} = 2.39 \text{ L } \times \frac{0.951 \text{ atm}}{1.316 \text{ atm}} = 1.727 = 1.73 \text{ L}$$

5.23   Using Boyle's law, let P_i = pressure of 345 cm^3 of gas, and solve for it:

$$P_i = P_f \times \frac{V_f}{V_i} = 2.51 \text{ kPa } \times \frac{0.0457 \text{ cm}^3}{315 \text{ cm}^3} = 3.641 \times 10^{-4} = 3.64 \times 10^{-4} \text{ kPa}$$

5.25   Use Charles's law: T_i = 18°C + 273 = 291 K, and T_f = 0°C + 273 = 273 K.

$$V_f = V_i \times \frac{T_f}{T_i} = 2.67 \text{ mL } \times \frac{273 \text{ K}}{291 \text{ K}} = 2.5048 = 2.50 \text{ mL}$$

5.27   Use the combined law: T_i = 26°C + 273 = 299 K, and T_f = 0°C + 273 = 273 K.

$$V_f = V_i = \times \frac{P_i}{P_f} \times \frac{T_f}{T_i} = 35.5 \text{ mL } \times \frac{741 \text{ mmHg}}{760 \text{ mmHg}} \times \frac{273 \text{ K}}{299 \text{ K}} = 31.60 = 31.6 \text{ mL}$$

5.29 The balanced equation is

$$4NH_3 + 5O_2 \rightarrow 4NO + 6H_2O$$

The ratio of moles of NH_3 to moles of NO = 4 to 4, or 1 to 1, so one volume of NH_3 will produce one volume of NO at the same temperature and pressure.

5.31 Solve the ideal gas law for V:

$$V = = \frac{nRT}{P} = nRT\left(\frac{1}{P}\right)$$

If the temperature and number of moles are held constant, then the product nRT is constant, and the volume is inversely proportional to the pressure:

$$V = \text{constant} \times \frac{1}{P}$$

5.33 Using the moles of chlorine, solve the ideal gas law for V:

$$V = \frac{nRT}{P} = \frac{(2.75 \text{ mol})(0.08206 \text{ L} \cdot \text{atm/K} \cdot \text{mol})(294 \text{ K})}{3.78 \text{ atm}} = 17.\underline{5}5 = 17.6 \text{ L}$$

5.35 Because density equals mass per unit volume, calculating the mass of 1 L (exact number) of gas will give the density of the gas. Start from the ideal gas law and calculate n; then convert the moles of gas to grams using the molar mass.

$$n = \frac{PV}{RT} = \frac{(751/760 \text{ atm})(1 \text{ L})}{(0.08206 \text{ L} \cdot \text{atm/K} \cdot \text{mol})(304 \text{ K})} = 0.039\underline{6}1 \text{ mol}$$

$$0.039\underline{6}1 \text{ mol} \times \frac{17.03 \text{ g}}{1 \text{ mol}} = 0.67\underline{4}58 = 0.675 \text{ g}$$

Therefore the density of NH_3 at 31°C is 0.675 g/L.

5.37   The ideal gas law gives *n* moles, which then is divided into the mass of 1.731 g for
        molar mass.

$$n = \frac{PV}{RT} = \frac{(753/760 \text{ atm}) (1 \text{ L})}{(0.08206 \text{ L} \cdot \text{atm/K} \cdot \text{mol}) (368 \text{ K})} = 0.03280 \text{ mol}$$

$$\text{Molar mass} = \frac{1.731 \text{ g}}{0.03280 \text{ mol}} = 52.75 \text{ g/mol}$$

The molecular weight is 52.8 amu.

5.39   For a gas at a given temperature and pressure, the density depends on molecular
        weight (or for a mixture, the average molecular weight). Thus, at the same temperature
        and pressure, the density of NH_4Cl gas would be greater than that of a mixture of NH_3
        and HCl because the average molecular weight of NH_3 and HCl would be lower than
        that of NH_4Cl.

5.41   The 0.0725 mol CaC_2 will form 0.0725 mol C_2H_2. The volume is found from the ideal
        gas law:

$$\text{Vol} = \frac{(0.0725 \text{ mol})(0.08206 \text{ L} \cdot \text{atm/K} \cdot \text{mol})(299 \text{ K})}{715/760 \text{ atm}} = 1.890 = 1.89 \text{ L}$$

5.43   Use the equation to obtain the moles of CO_2 and then the ideal gas law to obtain the
        volume.

$$2LiOH(s) + CO_2(g) \rightarrow Li_2CO_3(aq) + H_2O(l)$$

$$327 \text{ g LiOH} \times \frac{1 \text{ mol LiOH}}{23.95 \text{ g LiOH}} \times \frac{1 \text{ mol } CO_2}{2 \text{ mol LiOH}} = 6.8267 \text{ mol } CO_2$$

$$V = \frac{(6.8267 \text{ mol})(0.08206 \text{ L} \cdot \text{atm/K} \cdot \text{mol})(294 \text{ K})}{781/760 \text{ atm}} = 160.2 = 160. \text{ L}$$

5.45   Use the equation to obtain the moles of ammonia and then the ideal gas law to obtain the volume.

$$2NH_3(g) + H_2SO_4(aq) \rightarrow (NH_4)_2SO_4(aq)$$

$$150.0 \text{ g } (NH_4)_2SO_4 \times \frac{1 \text{ mol } (NH_4)_2SO_4}{132.1 \text{ g } (NH_4)_2SO_4} \times \frac{2 \text{ mol } NH_3}{1 \text{ mol } (NH_4)_2SO_4}$$

$$= 2.2710 \text{ mol } NH_3$$

$$V = \frac{nRT}{P} = \frac{(2.2710 \text{ mol})(0.08206 \text{ L} \cdot \text{atm/K} \cdot \text{mol})(291 \text{ K})}{0.915 \text{ atm}}$$

$$= 59.268 = 59.3 \text{ L}$$

5.47   For each gas, P(gas) = P x (mole fraction of gas).

$P(H_2)$ = 760 mmHg x 0.250   = 190.0   = 190 mmHg

$P(CO_2)$ = 760 mmHg x 0.650   = 494.0   = 494 mmHg

$P(HCl)$ = 760 mmHg x 0.054   = 41.04   = 41 mmHg

$P(HF)$ = 760 mmHg x 0.028   = 21.28   = 21 mmHg

$P(SO_2)$ = 760 mmHg x 0.017   = 12.92   = 13 mmHg

$P(H_2S)$ = 760 mmHg x 0.001   = 0.76   = 0.8 mmHg

5.49   The total pressure is the sum of the partial pressures of CO and H_2O, so

$$P_{CO} = P - P_{water} = 743 \text{ mmHg} - 23.8 \text{ mmHg} = 719.2 \text{ mmHg}$$

$$n_{CO} = \frac{P_{CO}V}{RT} = \frac{(719.2/760 \text{ atm})(2.68 \text{ L})}{(0.08206 \text{ L} \cdot \text{atm/K} \cdot \text{mol})(298 \text{ K})} = 0.1037 \text{ mol CO}$$

$$0.1037 \text{ mol CO} \times \frac{1 \text{ mol HCOOH}}{1 \text{ mol CO}} \times \frac{46.03 \text{ g HCOOH}}{1 \text{ mol HCOOH}}$$

$$= 4.773 = 4.77 \text{ g HCOOH}$$

5.51   Substitute 330 K (57°C) into Maxwell's distribution:

$$u_{330\,K} = \sqrt{\frac{3RT}{M_m}} = \sqrt{\frac{3 \times 8.31\ kg \cdot m^2/(s^2 \cdot K \cdot mol) \times 330\ K}{352 \times 10^{-3}\ kg/mol}}$$

$$= 1.5\underline{2}8 \times 10^2 = 1.53 \times 10^2\ m/s$$

5.53   Because $u(CO_2) = u(CH_4)$, we can equate the two right-hand sides of the Maxwell distributions:

$$\sqrt{\frac{3RT(CO_2)}{M_m(CO_2)}} = \sqrt{\frac{3RT(CH_4)}{M_m(CH_4)}}$$

Squaring both sides, rearranging to solve for $T(CO_2)$, and substituting numerical values, we have

$$T(CO_2) = T(CH_4) \times \frac{M_m(CO_2)}{M_m(CH_4)} = 298\ K \times \frac{44.01\ g/mol}{16.032\ g/mol} = 81\underline{8}.0\ K$$

Thus the temperature is

$$T = 81\underline{8}.0 - 273 = 54\underline{5}.0 = 545°C$$

5.55   Because the diffusion occurs at the same temperature, $T(gas) = T(Ar)$. A ratio of two Maxwell distributions can be written, but it can be simplified by canceling the 3 x 8.31 terms and rearranging the denominators. To simplify the definition of the rates, we assume the time is 1 second and define the rate(Ar) as 9.23 mL/1 s and the rate(gas) as 4.83 mL/1 s. Then we write a ratio of two Maxwell distributions:

$$\frac{u_{gas}}{u_{Ar}} = \frac{4.83\ mL/1\ s}{9.23\ mL/1\ s} = \frac{\sqrt{M_m(Ar)}}{\sqrt{M_m(gas)}} = \frac{\sqrt{39.95}}{\sqrt{M_m(gas)}}$$

Squaring both sides gives

$$0.27\underline{3}83 = \frac{39.95}{M_m(gas)}$$

Solving for the molar mass gives

$$M_m(gas) = \frac{39.95}{0.27383} = 14\underline{5}.8\ g/mol;\quad molecular\ weight = 146\ amu$$

5.57 Solving the van der Waals equation for $n = 1$ and $T = 355.2$ K for P gives

$$P = \frac{RT}{(V - b)} - \frac{a}{V^2} = \frac{(0.08206 \text{ L} \cdot \text{atm/K} \cdot \text{mol})(355.2 \text{ K})}{(30.00 \text{ L} - 0.08710 \text{ L})} - \frac{12.56 \text{ L}^2 \cdot \text{atm}}{(30.00 \text{ L})^2}$$

$P = 0.974\underline{4}19 - 0.0139\underline{5}55 = 0.960\underline{4}6 = 0.9605$ atm

P(ideal gas law) $= 0.971\underline{5}9$ atm

■ Solutions to General Problems

5.59 Calculate the mass of 1 cm^2 of the 20.5 m of water above the air in the glass. The volume is the product of the area of 1 cm^2 and the height of 20.5 x 10^2 cm (20.5 m) of water. The density of 1.00 g/cm^3 must be used to convert volume to mass:

$m = d \times V$

$m = 1.00$ g/cm^3 x (1.00 cm^2 x 20.5 x 10^2 cm) $= 2.05$ x 10^3 g, or 2.05 kg

The pressure exerted on an object at the bottom of the column of water is

$$P = \frac{\text{force}}{\text{area}} = \frac{(m)(g)}{\text{area}} = \frac{(2.05 \text{ kg})(9.807 \text{ m/s}^2)}{(1.00 \text{ cm}^2)\left(\dfrac{10^{-2} \text{ m}}{1 \text{ cm}}\right)^2} = 2.01 \times 10^5 \text{ kg/ms}^2$$

$$= 2.01 \times 10^5 \text{ Pa}$$

The total pressure on the air in the tumbler equals the barometric pressure and the water pressure:

$P = 1.00$ x 10^2 kPa + 2.01 x 10^2 kPa $= 3.01$ x 10^2 kPa

Multiply the initial volume by a factor accounting for the change in pressure to find V_f:

$$V_f = V_i \times \frac{P_i}{P_f} = 243 \text{ cm}^3 \times \left(\frac{1.00 \times 10^2 \text{ kPa}}{3.01 \times 10^2 \text{ kPa}}\right) = 80.\underline{7}3 = 80.7 \text{ cm}^3$$

5.61 Use the combined gas law and solve for V_f:

$$V_f = V_i \times \frac{P_i}{P_f} \times \frac{T_f}{T_i} = 5.0 \text{ dm}^3 \times \frac{100.0 \text{ kPa}}{79.0 \text{ kPa}} \times \frac{293 \text{ K}}{287 \text{ K}} = 6.\underline{4}6 = 6.5 \text{ dm}^3$$

5.63 Use the ideal gas law to calculate the moles of helium and combine this with Avogadro's number to obtain the number of helium atoms:

$$n = \frac{PV}{RT} = \frac{(742/760 \text{ atm})(0.00831 \text{ L})}{(0.08206 \text{ L} \cdot \text{atm/K} \cdot \text{mol})(296 \text{ K})} = 3.3401 \times 10^{-4} \text{ mol}$$

$$3.3401 \times 10^{-4} \text{ mol He} \times \frac{6.022 \times 10^{23} \text{ He}^{2+} \text{ ions}}{1 \text{ mol He}} \times \frac{1 \text{ atom}}{1 \text{ He}^{2+} \text{ ion}} = 2.011 \times 10^{21}$$

$$= 2.01 \times 10^{20} \text{ He atoms}$$

5.65 Use the ideal gas law to calculate the moles of CO_2. Then convert to mass of LiOH.

$$n = \frac{PV}{RT} = \frac{(1.00 \text{ atm})(5.8 \times 10^2 \text{ L})}{(0.08206 \text{ L} \cdot \text{atm/K} \cdot \text{mol})(273 \text{ K})} = 25.89 \text{ mol } CO_2$$

$$25.89 \text{ mol } CO_2 \times \frac{2 \text{ mol LiOH}}{1 \text{ mol } CO_2} \times \frac{23.95 \text{ g LiOH}}{1 \text{ mol LiOH}} = 1240 = 1.2 \times 10^3 \text{ g LiOH}$$

5.67 Convert mass to moles of $KClO_3$, and then use the equation below to convert to moles of O_2. Use the ideal gas law to convert moles of O_2 to pressure at 25°C (298 K).

$$2KClO_3(s) \rightarrow 2KCl(s) + 3O_2(g)$$

$$170.0 \text{ g } KClO_3 \times \frac{1 \text{ mol } KClO_3}{122.55 \text{ g } KClO_3} \times \frac{3 \text{ mol } O_2}{2 \text{ mol } KClO_3} = 2.080 \text{ mol } O_2$$

$$P = \frac{PV}{RT} = \frac{(2.080 \text{ mol})(0.08206 \text{ L} \cdot \text{atm/ K} \cdot \text{mol})(298 \text{ K})}{2.50 \text{ L}}$$

$$= 20.353 = 20.4 \text{ atm } O_2$$

5.69 The number of moles of carbon dioxide is:

$$n = \frac{646 \ / 760 \text{ atm} \times 0.1500 \text{ L}}{0.08206 \text{ L} \cdot \text{at m/ K} \cdot \text{mol} \times 300 \text{ K}} = 0.005179 \text{ mol}$$

The number of moles of molecular acid used is:

$$0.1250 \text{ mol /L} \times 0.04141 \text{ L} = 0.005176 \text{ mol acid}$$

Thus the acid is H_2SO_4 since one mole of H_2SO_4 reacts to form one mole of CO_2

5.71 Use Maxwell's distribution to calculate the temperature in kelvins; then convert to °C.

$$T = \frac{u^2 M_m}{3R} = \frac{(0.49 \times 10^3 \text{ m/s})^2 (30.048 \times 10^{-3} \text{ kg/mol})}{3 (8.31 \text{ kg} \cdot \text{m}^2/\text{s}^2 \cdot \text{K} \cdot \text{mol})}$$

$$= 28\underline{9}.3 = 2.9 \times 10^2 \text{ K } (16 \text{ °C})$$

5.73 Use $CO + 1/2 O_2 \rightarrow CO_2$, instead of 2CO. First find the moles of CO and O_2 by using the ideal gas law.

$$n_{CO} = \frac{PV}{RT} = \frac{(0.500 \text{ atm}) (2.00 \text{ L})}{(0.08206 \text{ L} \cdot \text{atm/K} \cdot \text{mol}) (300 \text{ K})} = 0.040\underline{6}2 \text{ mol}$$

$$n_{O_2} = \frac{PV}{RT} = \frac{(1.00 \text{ atm}) (1.00 \text{ L})}{(0.08206 \text{ L} \cdot \text{atm/K} \cdot \text{mol}) (300 \text{ K})} = 0.040\underline{6}2 \text{ mol}$$

There are equal amounts of CO and O_2, but (from the equation) only half as many moles of O_2 as CO are required for the reaction. Therefore, when 0.04062 moles of CO have been consumed, only 0.04062/2 moles of O_2 will have been used up. Then 0.04062/2 mol O_2 will remain, and 0.04062 mol CO_2 will have been produced. At the end,

$$n_{CO} = 0 \text{ mol}; \quad n_{O_2} = 0.0202 \text{ mol}; \text{ and } n_{CO_2} = 0.04062 \text{ mol}$$

However, the total volume with the valve open is 3.00 L, so the partial pressures of O_2 and CO_2 must be calculated from the ideal gas law for each:

$$\frac{nRT}{V} = \frac{(0.0203 \text{ mol } O_2) (0.08206 \text{ L} \cdot \text{atm/K} \cdot \text{mol}) (300 \text{ K})}{3.00 \text{ L}}$$

$$= 0.166\underline{5}8 = 0.167 \text{ atm } O_2$$

$$\frac{nRT}{V} = \frac{(0.04062 \text{ mol } CO_2) (0.08206 \text{ L} \cdot \text{atm/K} \cdot \text{mol}) (300 \text{ K})}{3.00 \text{ L}}$$

$$= 0.333\underline{3}2 = 0.333 \text{ atm } CO_2$$

■ Solutions to Cumulative-Skills Problems

5.75   Assume a 100.0-g sample, giving 85.2 g CH_4 and 14.8 g C_2H_6. Convert each to moles:

$$85.2 \text{ g } CH_4 \times \frac{1 \text{ mol } CH_4}{16.04 \text{ g } CH_4} = 5.3\underline{1}1 \text{ mol } CH_4$$

$$14.8 \text{ g } C_2H_6 \times \frac{1 \text{ mol } CH_4}{16.04 \text{ g } CH_4} = 0.49\underline{2}1 \text{ mol } C_2H_6$$

$$V_{CH_4} = \frac{(5.311 \text{ mol})(0.08206 \text{ L} \cdot \text{atm/K} \cdot \text{mol})(291 \text{ K})}{(771/760) \text{ atm}} = 12\underline{5}.03 \text{ L}$$

$$V_{C_2H_6} = \frac{(0.4921 \text{ mol})(0.08206 \text{ L} \cdot \text{atm/K} \cdot \text{mol})(291 \text{ K})}{(771/760) \text{ atm}} = 11.\underline{5}8 \text{ L}$$

The density is calculated as follows:

$$d = \frac{85.2 \text{ g } CH_4 + 14.8 \text{ g } C_2H_6}{(125.03 + 11.58) \text{ L}} = 0.73\underline{1}9 = 0.732 \text{ g/L}$$

5.77   First subtract the height of mercury equivalent to the 25.00 cm (250 mm) of water inside the tube from 771 mmHg to get P_{gas}. Then subtract the vapor pressure of water, 18.7 mmHg, from P_{gas} to get P_{O_2}

$$h_{Hg} = \frac{(h_W)(d_W)}{d_{Hg}} = \frac{250 \text{ mm} \times 0.99987 \text{ g/cm}^3}{13.596 \text{ g/cm}^3}$$

$$P_{gas} = P - P_{25 \text{ cm water}} = 771 \text{ mmHg} - 18.38 \text{ mmHg} = 75\underline{2}.62 \text{ mmHg}$$

$$P_{O_2} = 75\underline{2}.62 \text{ mmHg} - 18.7 \text{ mmHg} = 73\underline{3}.92 \text{ mmHg}$$

$$n = \frac{PV}{RT} = \frac{(733.92/760 \text{ atm})(0.0310 \text{ L})}{(0.08206 \text{ L} \cdot \text{atm/K} \cdot \text{mol})(294 \text{ K})} = 0.00124\underline{0}8 \text{ mol } O_2$$

$$\text{Mass} = (2 \times 0.0012408) \text{ mol } Na_2O_2 \times \frac{77.98 \text{ g } Na_2O_2}{1 \text{ mol } Na_2O_2}$$

$$= 0.19\underline{3}5 = 0.194 \text{ g } Na_2O_2$$

5.79   First find the moles of CO_2:

$$n = \frac{PV}{RT} = \frac{(785/760 \text{ atm}) (1.94 \text{ L})}{(0.08206 \text{ L} \bullet \text{atm/K} \bullet \text{mol}) (298 \text{ K})} = 0.08194 \text{ mol CO}$$

Set up one equation in one unknown: x = mol $CaCO_3$; (0.08194 - x) = mol $MgCO_3$.

7.85 g = (100.1 g/mol)x + (84.32 g/mol)(0.08194 - x)

$$x = \frac{(7.85 - 6.9092)}{(100.1 - 84.32)} = 0.05962 \text{ mol CaCO}_3$$

(0.08194 - x) = 0.02232 mol $MgCO_3$

$$CaCO_3 = \frac{0.05962 \text{ mol CaCO}_3 \times 100.1 \text{ g/mol}}{7.85 \text{ g}} \times 100\% = 76.02 = 76\%$$

% $MgCO_3$ = 100.00% - 76.02% = 23.98 = 24%

6. THERMOCHEMISTRY

■ Answers to Review Questions

6.1   Energy is the potential or capacity to move matter. Kinetic energy is the energy
       associated with an object by virtue of its motion. Potential energy is the energy an
       object has by virtue of its position in a field of force. Internal energy is the sum of the
       kinetic and potential energies of the particles making up a substance.

6.2   In terms of SI base units, a joule is $kg \cdot m^2/s^2$. This is equivalent to the units for the
       kinetic energy ($\frac{1}{2}mv^2$) of an object of mass m (in kg) moving with speed v (in m/s).

6.3   Originally, a calorie was defined as the amount of energy required to raise the
       temperature of one gram of water by one degree Celsius. Now the calorie is defined as
       4.184 J.

6.4   At either of the two highest points above the earth in a pendulum's cycle, the energy of
       the pendulum is all potential energy and is equal to the product mgh (m = mass of
       pendulum, g = constant acceleration of gravity, and h = height of pendulum). As the
       pendulum moves downward, its potential energy decreases from mgh to near zero,
       depending on how close it comes to the earth's surface. During the downward motion,
       its potential energy is converted to kinetic energy. When it reaches the lowest point
       (middle) of its cycle, the pendulum has its maximum kinetic energy and minimum
       potential energy. As it rises above the lowest point, its kinetic energy begins to be
       converted to potential energy. When it reaches the other high point in its cycle, the
       energy of the pendulum is again all potential energy. By the law of conservation of
       energy, this energy cannot be lost, only converted to other forms. At rest, the energy of
       the pendulum has been transferred to the surroundings in the form of heat.

6.5   As the heat flows into the gas, the gas molecules gain energy and move at a faster
       average speed. The internal energy of the gas increases.

6.6   An exothermic reaction is a chemical reaction or a physical change in which heat is evolved (q is negative). For example, burning one mol of methane, $CH_4(g)$, yields carbon dioxide, water, and 890.3 kJ of heat. An endothermic reaction is a chemical reaction or physical change in which heat is absorbed (q is positive). For example, the reaction of one mol of barium hydroxide with ammonium nitrate absorbs 170.8 kJ of heat in order to form ammonia, water, and barium nitrate.

6.7   Changes in internal energy depend only on the initial and final states of the system, which are determined by variables such as temperature and pressure. Such changes do not depend on any previous history of the system.

6.8   It is important to give the states when writing an equation for ΔH because ΔH depends on the states of all reactants and products. If any state changes, ΔH changes.

6.9   First convert the 10.0 g of water to moles of water, using its molar mass (18.02 g/mol). Next, using the equation, multiply the moles of water by the appropriate mole ratio (1 mol CH_4 / 2 mol H_2O). Finally, multiply the moles of CH_4 by the heat of the reaction (-890.3 kJ/mol CH_4).

6.10   The heat capacity (C) of a substance is the quantity of heat needed to raise the temperature of the sample of substance one degree Celsius (or one kelvin). The specific heat of a substance is the quantity of heat required to raise the temperature of one gram of a substance by one degree Celsius (or one kelvin) at constant pressure.

6.11   Hess's law states that, for a chemical equation that can be written as the sum of two or more steps, the enthalpy change for the overall equation is the sum of the enthalpy changes for the individual steps. In other words, no matter how you go from reactants to products, the enthalpy change for the overall chemical change is the same. This is because enthalpy is a state function.

6.12   The thermodynamic standard state refers to the standard thermodynamic conditions chosen for substances when listing or comparing thermodynamic data: one atm pressure and the specified temperature (usually 25°C).

6.13   The reference form of an element is the most stable form (physical state and allotrope) of the element under standard thermodynamic conditions. The standard enthalpy of formation of an element in its reference form is zero.

6.14   The standard enthalpy of formation of a substance, $\Delta H°_f$, is the enthalpy change for the formation of one mole of the substance in its standard state from its elements in their reference form and in their standard states.

6.15   A fuel is any substance that is burned or similarly reacted to provide heat and other forms of energy. The fossil fuels are petroleum (oil), gas, and coal. They were formed millions of years ago when aquatic plants and animals were buried and compressed by layers of sediment at the bottoms of swamps and seas. Over time this organic matter was converted by bacterial decay and pressure to fossil fuels.

6.16   One of the ways of converting coal to methane involves the water-gas reaction.

$$C(s) + H_2O(g) \rightarrow CO(g) + H_2(g)$$

In this reaction, steam is passed over hot coal. This mixture is then reacted over a catalyst to give methane.

$$CO(g) + 3H_2(g) \rightarrow CH_4(g) + H_2O(g)$$

■ Solutions to Practice Problems

Note on significant figures: If the final answer to a solution needs to be rounded off, it is given first with one nonsignificant figure, and the last significant figure is underlined. The final answer is then rounded to the correct number of significant figures. In multiple-step problems, intermediate answers are given with at least one nonsignificant figure; however, only the final answer has been rounded off.

6.23   The kinetic energy, in J, is

$$E_k = 1/2 \times 3.15 \times 10^3 \text{ lb} \times \frac{0.4536 \text{ kg}}{1 \text{ lb}} \times \left[\frac{65 \text{ mi}}{1 \text{ h}}\right]^2 \times \left[\frac{1609 \text{ m}}{1 \text{ mi}}\right]^2 \times \left[\frac{1 \text{ h}}{3600 \text{ s}}\right]^2$$

$$= 6.0\underline{2} \times 10^5 = 6.0 \times 10^5 \text{ J}$$

The kinetic energy, in calories, is

$$E_k = 6.02 \times 10^5 \text{ J} \times \frac{1 \text{ cal}}{4.184 \text{ J}} = 1.4\underline{4} \times 10^5 = 1.4 \times 10^5 \text{ cal}$$

6.25   To insert the mass of one molecule of ClO_2 in the formula, multiply the molar mass by the reciprocal of Avogadro's number. The kinetic energy in J is

$$E_k = 1/2 \times \frac{67.45 \text{ g}}{1 \text{ mol}} \times \frac{1 \text{ mol}}{6.022 \times 10^{23} \text{ molec.}} \times \frac{1 \text{ kg}}{1000 \text{ g}} \times \left[\frac{306 \text{ m}}{1 \text{ s}}\right]^2$$

$$= 5.2\underline{4}3 \times 10^{-21} = 5.24 \times 10^{-21} \text{ J/molec.}$$

6.27 The gain of 66.2 kJ of heat per 2 mol NO_2 means the reaction is endothermic. Because energy is gained by the system from the surroundings, q is positive, and because the reaction involves 2 mol NO_2, q for the reaction is +66.2 kJ.

6.29 The reaction of Fe(s) with HCl must yield H_2 and $FeCl_3$. To balance the hydrogen, 2HCl must be written first as a reactant. However, to balance the chlorine, 3 x 2HCl must be written finally as a reactant, and $2FeCl_3$ must be written as a product:

$$2Fe(s) + 6HCl(aq) \rightarrow 2FeCl_3(aq) + 3H_2(g)$$

To write a thermochemical equation, the sign of ΔH must be negative because heat is evolved:

$$2Fe(s) + 6HCl(aq) \rightarrow 2FeCl_3(aq) + 3H_2(g); \quad \Delta H = 2 \times -89.1 \text{ kJ} = -178.2 \text{ kJ}$$

6.31 The first equation is

$$P_4(s) + 5O_2(g) \rightarrow P_4O_{10}(s); \quad \Delta H = -3010 \text{ kJ}$$

The second equation is

$$P_4O_{10}(s) \rightarrow P_4(s) + 5O_2(g); \quad \Delta H = ?$$

The second equation has been obtained by reversing the first equation. Therefore, to obtain ΔH for the second equation, ΔH for the first equation must be reversed in sign:

$$-(-3010) = +3010 \text{ kJ} = 3.010 \times 10^3 \text{ kJ}$$

6.33 Because nitrogen dioxide is written as NO_2 in the equation, the molar mass of NO_2 = 46.01 g per mol NO_2. From the equation, 2 mol NO_2 evolve 114 kJ heat. Divide the 114 kJ by the 46.01 g/mol NO_2 to obtain the amount of heat evolved per gram of NO_2:

$$\frac{-114 \text{ kJ}}{2 \text{ mol NO}_2} \times \frac{1 \text{ mol NO}_2}{46.01 \text{ g NO}_2} = -1.2\underline{3}8 = -1.24 \text{ kg/g NO}_2$$

6.35 The molar mass of C_3H_8 is 44.06 g/mol. From the equation, 1 mol C_3H_8 evolves 2044 kJ heat. Divide 425 kJ by 2044 kJ, and multiply by the molar mass to obtain the mass needed:

$$-425 \text{ kJ} \times \frac{1 \text{ mol C}_3\text{H}_8}{-2044 \text{ kJ}} \times \frac{44.06 \text{ g C}_3\text{H}_8}{1 \text{ mol C}_3\text{H}_8} = 9.1\underline{6}1 = 9.16 \text{ g C}_3\text{H}_8$$

6.37   Multiply the 180 g (0.180 kg) of water by the specific heat of 4.18 J/(g•°C) and by Δt to obtain heat in joules:

$$180 \text{ g} \times (96°C - 21°C) \times \frac{4.18 \text{ J}}{1 \text{ g} \cdot °C} = 5\underline{6}430 = 5.6 \times 10^4 \text{ J}$$

6.39   The enthalpy change for the reaction is equal in magnitude and opposite in sign to the heat-energy change occurring from the cooling of the solution and calorimeter.

$$q_{calorimeter} = (1071 \text{ J/°C})(21.56°C - 25.00°C) = -36\underline{8}4.2 \text{ J}$$

Thus 15.3 g $NaNO_3$ is equivalent to -3684.2 J heat energy. The amount of heat absorbed by 1.000 mol $NaNO_3$ is calculated from +3684.2 J (opposite sign):

$$1.000 \text{ mol NaNO}_3 \times \frac{85.00 \text{ g NaNO}_3}{1 \text{ mol NaNO}_3} \times \frac{3684.2 \text{ J}}{15.3 \text{ g NaNO}_3} = 2.0\underline{4}68 \times 10^4 \text{ J}$$

Thus, the enthalpy change, ΔH, for the reaction is 2.05 x 10^4 J, or 20.5 kJ, per mol $NaNO_3$.

6.41   The energy change for the reaction is equal in magnitude and opposite in sign to the heat energy produced from the warming of the solution and the calorimeter.

$$q_{calorimeter} = (9.63 \text{ kJ/°C})(33.73 - 25.00°C) = +84.\underline{0}69 \text{ kJ}$$

Thus 2.84 g C_2H_5OH is equivalent to 84.069 kJ heat energy. The amount of heat released by 1.000 mol C_2H_5OH is calculated from -84.069 kJ (opposite sign):

$$1.00 \text{ mol C}_2\text{H}_5\text{OH} \times \frac{46.07 \text{ g C}_2\text{H}_5\text{OH}}{1 \text{ mol C}_2\text{H}_5\text{OH}} \times \frac{-84.069 \text{ kJ}}{2.84 \text{ g C}_2\text{H}_5\text{OH}} = -13\underline{6}3.75 \text{ kJ}$$

Thus the enthalpy change ΔH for the reaction is -1.36 x 10^3 kJ/mol ethanol.

6.43   Using the equations in the data, reverse the direction of the first reaction, and reverse the sign of its ΔH. Then multiply the second equation by two, multiply its ΔH by two, and add. Setup:

$N_2(g)$ +	$2H_2O(l)$	$\rightarrow$	$N_2H_4(l)$ +	$O_2(g)$;	ΔH = (-622.2 kJ) x (-1)
$2H_2(g)$ +	$O_2(g)$	$\rightarrow$	$2H_2O(l)$;		ΔH = (-285.8 kJ) x (2)
$N_2(g)$ +	$2H_2(g)$	$\rightarrow$	$N_2H_4(l)$;		ΔH = 50.6 kJ

6.45   After reversing the first equation in the data, add all the equations. Setup:

$$C_2H_4(g) + 3O_2(g) \rightarrow 2CO_2(g) + 2H_2O(l); \quad \Delta H = (-1411 \text{ kJ})$$

$$2CO_2(g) + 3H_2O(l) \rightarrow C_2H_6(g) + 7/2O_2(g); \quad \Delta H = (-1560) \times (-1)$$

$$H_2(g) + 1/2O_2(g) \rightarrow H_2O(l); \quad \Delta H = (-286 \text{ kJ})$$

$$C_2H_4(g) + H_2(g) \rightarrow C_2H_6(g); \quad \Delta H = -137 \text{ kJ}$$

6.47   Write the $\Delta H°$ values (Appendix B) underneath each compound in the balanced equation:

$$C_2H_5OH(l) \rightarrow C_2H_5OH(g)$$
$$\text{-277.7} \qquad \text{-235.1} \qquad \text{(kJ)}$$

$$\Delta H°_{vap} = [\Delta H°(C_2H_5OH)(g)] - [\Delta H°(C_2H_5OH)(l)] = [-235.1] - [-277.7] = +42.6 \text{ kJ}$$

6.49   Write the $\Delta H°$ values (Appendix B) underneath each compound in the balanced equation:

$$Fe_3O_4(s) + 4CO(g) \rightarrow 3Fe(s) + 4CO_2(g)$$
$$\text{-1120.9} \qquad 4(\text{-110.5}) \qquad 3(0) \qquad 4(\text{-393.5}) \qquad \text{(kJ)}$$

$$\Delta H° = \Sigma n\Delta H°(\text{products}) - \Sigma m\Delta H°(\text{reactants})$$

$$= [3(0) + 4(-393.5)] - [(-1120.9) + 4(-110.5)] = -11.1 \text{ kJ}$$

6.51   Calculate the molar heat of formation from the equation with the $\Delta H°_f$ values below each substance; then convert to the heat for 10.0 g of $MgCO_3$ using the molar mass of 84.3.

$$MgCO_3(s) \rightarrow MgO(s) + CO_2(g)$$
$$\text{-1111.7} \qquad \text{-601.2} \qquad \text{-393.5} \qquad \text{kJ}$$

$$\Delta H° = \Sigma n\Delta H°(\text{products}) - \Sigma m\Delta H°(\text{reactants})$$

$$\Delta H° = [-601.2 \text{ kJ} + (-393.5 \text{ kJ})] - (-1111.7 \text{ kJ}) = 117.0 \text{ kJ}$$

$$\text{Heat} = 25.0 \text{ g} \times \frac{1 \text{ mol}}{84.3 \text{ g}} \times \frac{117.0 \text{ kJ}}{\text{mol}} = 34.69 = 34.7 \text{ kJ}$$

■ Solutions to General Problems

6.53 Using Table 1.4 and 4.184 J/cal, convert the 686 Btu/lb to J/g:

$$\frac{686\ Btu}{1\ lb} \times \frac{252\ cal}{1\ Btu} \times \frac{4.184\ J}{1\ cal} \times \frac{1\ lb}{0.4536\ kg} \times \frac{1\ kg}{10^3\ g}$$

$$= 1.5\underline{9}4 \times 10^3 = 1.59 \times 10^3\ J/g$$

6.55 The equation is

$$CaCO_3(s) \rightarrow CaO(s) + CO_2(g);\quad \Delta H = 177.9\ kJ$$

Use the molar mass of 100.08 g/mol to convert the heat per mol to heat per 45.6 g.

$$45.6\ g\ CaCO_3 \times \frac{1\ mol\ CaCO_3}{100.08\ g\ CaCO_3} \times \frac{177.9\ kJ}{1\ mol\ CaCO_3} = 81.\underline{0}5 = 81.1\ kJ$$

6.57 The equation is

$$2HCHO_2(l) + O_2(g) \rightarrow 2CO_2(g) + 2H_2O(l)$$

Use the molar mass of 46.03 g/mol to convert -30.3 kJ/5.48 g to ΔH per mol of acid.

$$\frac{-30.3\ kJ}{5.48\ g\ HCHO_2} \times \frac{46.03\ g\ HCHO_2}{1\ mol\ HCHO_2} = -25\underline{4}.508 = -255\ kJ/mol$$

6.59 Divide the 235 J heat by the mass of lead and the Δt to obtain the specific heat.

$$Specific\ heat = \frac{235\ J}{121.6\ g\ (35.5°C\ -\ 20.4°C)} = 0.127\underline{9}8 = \frac{0.128\ J}{g \cdot °C}$$

6.61 Use Δt and the heat capacity of 547 J/°C to calculate q:

$$q = C\Delta t = (547\ J/°C)(36.66\ -\ 25.00)°C = 6.3\underline{7}8 \times 10^3\ J\ (6.3\underline{7}8\ kJ)$$

Energy is released in the solution process in raising the temperature, so ΔH is negative:

$$\Delta H = \frac{-6.378\ kJ}{6.48\ g\ LiOH} \times \frac{23.95\ g\ LiOH}{1\ mol\ LiOH} = -23.\underline{5}7 = -23.6\ kJ/mol$$

6.63 Use Δt and the heat capacity of 13.43 kJ/°C to calculate q:

$$q = C\Delta t = (13.43 \text{ kJ/°C})(35.81 - 25.00)°C = 145.\underline{1}78 \text{ kJ}$$

As in the previous two problems, the sign of ΔH must be reversed, making the heat negative:

$$\Delta H = \frac{-145.178 \text{ kJ}}{10.00 \text{ g HC}_2\text{H}_3\text{O}_2} \times \frac{60.05 \text{ g HC}_2\text{H}_3\text{O}_2}{1 \text{ mol HC}_2\text{H}_3\text{O}_2} = -871.\underline{7}9 = -871.8 \text{ kJ/mol}$$

6.65 Using the equations in the data, reverse the direction of the first reaction, and reverse the sign of its ΔH. Then add the second and third equations and their ΔH's.

$$H_2O(g) + SO_2(g) \rightarrow H_2S(g) + 3/2O_2(g); \quad \Delta H = (-519 \text{ kJ}) \times (-1)$$
$$H_2(g) + 1/2O_2(g) \rightarrow H_2O(g); \quad \Delta H = (-242 \text{ kJ})$$
$$1/8S_8(\text{rh.}) + O_2(g) \rightarrow SO_2(g); \quad \Delta H = (-297 \text{ kJ})$$

$$H_2(g) + 1/8S_8(\text{rh.}) \rightarrow H_2S(g); \quad \Delta H = -20 \text{ kJ}$$

6.67 Write the $\Delta H°$ values (Table 6.2) underneath each compound in the balanced equation. The $\Delta H_f°$ of -635 kJ/mol CaO is given in the problem.

$$CaCO_3(s) \rightarrow CaO(s) + CO_2(g)$$

| -1206.9 | -635.1 | -393.5 | (kJ) |

$$\Delta H° = [(-635.1) + (-393.5)] - [-1206.9] = 178.3 \text{ kJ}$$

6.69 Calculate the molar heat of reaction from the equation with the $\Delta H_f°$ values below each substance; then convert to the heat for the reaction at 25°C.

$$2H_2(g) + O_2(g) \rightarrow 2H_2O(l)$$

0.0 kJ   0.0 kJ   -285.8 kJ

$$\Delta H° = 2 \times (-285.8 \text{ kJ}) - 0 - 0 = -571.6 \text{ kJ}$$

The moles of oxygen in 3.50 L with a density of 1.11 g/L is:

$$3.50 \text{ L O}_2 \times \frac{1.11 \text{ g O}_2}{\text{L O}_2} \times \frac{1 \text{ mol O}_2}{32.0 \text{ g O}_2} = 0.12\underline{1}4 \text{ mol O}_2$$

(continued)

The heat of reaction from 0.1214 mol O_2 is:

$$0.1214 \text{ mol } O_2 \times \frac{-571.6 \text{ kJ}}{\text{mol } O_2} = -69.\underline{3}9 = -69.4 \text{ kJ}$$

6.71 First calculate the heat evolved from the molar amounts represented by the balanced equation. From Appendix B, obtain the individual heats of formation, and write those below the reactants and products in the balanced equation. Then multiply the molar heats of formation by the number of moles in the balanced equation, and write those products below the molar heats of formation.

$2Al(s)$ +	$3NH_4NO_3(s)$	$\rightarrow$	$3N_2(g)$ +	$6H_2O(g)$ +	$Al_2O_3(s)$	
0.0 kJ	-365.6 kJ		0.0 kJ	-241.8	-1675.7 kJ	(kJ/mol)
0.0 kJ	-1096.8 kJ		0.0 kJ	-1450.8	-1675.7 kJ	(kJ/eqn.)

The total heat of reaction of two moles of Al and three moles of NH_4NO_3 is

$$\Delta H = [-1450.8 - 1675.7] - (-1096.8 \text{ kJ}) = -2029.7 \text{ kJ}$$

Now 325 kJ represents the following fraction of the total heat of reaction:

$$\frac{325 \text{ kJ}}{2029.7 \text{ kJ}} = 0.16\underline{0}1$$

Thus 325 kJ requires the fraction 0.1601 of the moles of each reactant: 0.1601 x 2, or 0.3202, mol of Al and 0.1601 x 3, or 0.4804, mol of NH_4NO_3. The mass of each reactant and the mass of the mixture are as follows:

0.3202 mol Al x 26.98 g/mol Al = 8.6$\underline{4}$0 g of Al

0.4808 mol NH_4NO_3 x 80.04 g/mol NH_4NO_3 = 38.$\underline{4}$4 g NH_4NO_3

8.640 g Al + 38.44 g NH_4NO_3 = 47.$\underline{0}$8 = 47.1 g of the mixture

■ Solutions to Cumulative-Skills Problems

6.73   The heat lost, q, by the water(s) with a temperature higher than the final temperature must be equal to the heat gained, q, by the water(s) with a temperature lower than the final temperature.

   q(lost by water at higher temp) = q(gained by water at lower temp)

   $\Sigma\,(s \times m \times \Delta t) = \Sigma\,(s \times m \times \Delta t)$

Divide both sides of the equation by the specific heat, s, to eliminate this term, and substitute the other values. Since the mass of the water at 50.0°C is greater than the sum of the masses of the other two waters, assume the final temperature will be greater than 37°C and greater than 15°C.

Use this to set up the three Δt expressions, and write one equation in one unknown, letting t equal the final temperature. Simplify by omitting the "grams" from 45.0 g, 25.0 g, and 15.0 g.

   $\Sigma\,(m \times \Delta t) = \Sigma\,(m \times \Delta t)$

   45.0 x (50.0°C - t) = 25.0 x (t - 15.0°C) + 15.0 x (t - 37°C)

   2250°C - 45.0 t = 25.0 t - 375°C + 15.0 t - 555.0°C

   -85.0 t = -3180°C

   t = 37.$\underline{4}$1 = 37.4°C (the final temperature)

6.75   The equation is

   $4NH_3(g) + 5O_2(g) \rightarrow 4NO(g) + 6H_2O(g);\ \ \Delta H° = $ -906 kJ

First determine the limiting reactant by calculating the moles of NH_3 and of O_2; then, assuming one of the reactants is totally consumed, calculate the moles of the other reactant needed for the reaction.

   $10.0\ g\ NH_3 \times \dfrac{1\ mol\ NH_3}{17.03\ g\ NH_3} = 0.58\underline{7}2\ mol\ NH_3$

   $20.0\ g\ O_2 \times \dfrac{1\ mol\ O_2}{32.00\ g\ O_2} = 0.62\underline{5}0\ mol\ O_2$

(continued)

$$0.62\underline{5}0 \text{ mol } O_2 \ \times \ \frac{1 \text{ mol } O_2}{32.00 \text{ g } O_2} = 0.500 \text{ mol } NH_3 \text{ needed}$$

Because NH_3 is present in excess of what is needed, O_2 must be the limiting reactant. Now calculate the heat released on the basis of the complete reaction of 0.622250 mol O_2:

$$\Delta H = \frac{-906 \text{ kJ}}{5 \text{ mol } O_2} \ \times \ 0.6250 \text{ mol } O_2 = -11\underline{3}.25 = -113 \text{ kJ}$$

The heat released by the complete reaction of the 20.0 g (0.6250 mol) of $O_2(g)$ is 113 kJ.

6.77   The equation is

$$N_2(g) + 3H_2(g) \ \rightarrow \ 2NH_3(g); \ \ \Delta H° = -91.8 \text{ kJ}$$

a. To find the heat evolved from the production of 1.00 L of NH_3, convert the 1.00 L to mol NH_3, using the density and molar mass (17.03 g/mol). Then convert the moles to heat (ΔH) using $\Delta H°$:

$$\text{moles } NH_3 = \frac{d \times V}{Mwt} = \frac{0.696 \text{ g/L } \times 1 \text{ L}}{17.03 \text{ g/mol}} = 0.040\underline{8}7 \text{ mol } NH_3$$

$$\Delta H = \frac{-91.8 \text{ kJ}}{2 \text{ mol } NH_3} \ \times \ 0.040\underline{8}7 \text{ mol } NH_3 = -1.8\underline{7}6 \ (1.88 \text{ kJ heat evolved})$$

b. First find the moles of N_2 using the density and molar mass (28.02 g/mol). Then convert to the heat needed to raise the N_2 from 25°C to 400°C:

$$\text{moles } N_2 = \frac{d \times V}{Mwt} = \frac{1.145 \text{ g/L } \times 0.500 \text{ L}}{28.02 \text{ g/mol}} = 0.020\underline{4}3 \text{ mol } N_2$$

$$0.020\underline{4}3 \text{ mol } N_2 \ \times \ \frac{29.12 \text{ J}}{\text{mol} \cdot °C} \ \times \ (400 - 25)°C = 22\underline{3}.2 \text{ J} \ (0.22\underline{3}2 \text{ kJ})$$

$$\% \text{ heat for } N_2 = \frac{0.2232 \text{ kJ}}{1.876 \text{ kJ}} \ \times \ 100\% = 11.\underline{8}9 = 11.9\%$$

6.79   The glucose equation is

$$C_6H_{12}O_6 + 6O_2 \rightarrow 6CO_2 + 6H_2O; \quad \Delta H° = -2802.8 \text{ kJ}$$

Convert the 2.50×10^3 kcal to mol of glucose, using the $\Delta H°$ of -2802.8 kJ for the reaction and the conversion factor of 4.184 kJ/kcal:

$$2.50 \times 10^3 \text{ kcal} \times \frac{4.184 \text{ kJ}}{1.000 \text{ kcal}} \times \frac{1 \text{ mol glucose}}{2802.8 \text{ kJ}} = 3.7\underline{3}19 \text{ mol glucose}$$

Next convert mol glucose to mol LiOH using the above equation for glucose and the equation for LiOH:

$$2\text{LiOH}(s) + CO_2(g) \rightarrow Li_2CO_3(s) + H_2O(l)$$

$$3.7\underline{3}19 \text{ mol glucose} \times \frac{6 \text{ mol } CO_2}{1 \text{ mol glucose}} \times \frac{2 \text{ mol LiOH}}{1 \text{ mol } CO_2} = 44.\underline{7}83 \text{ mol LiOH}$$

Finally, use the molar mass of LiOH to convert moles to mass:

$$44.783 \text{ mol LiOH} \times \frac{23.95 \text{ g LiOH}}{1 \text{ mol LiOH}} = 1.0\underline{7}25 \times 10^3 \text{ g } (1.07 \text{ kg}) \text{ LiOH}$$

7.   QUANTUM THEORY OF THE ATOM

■ Answers to Review Questions

7.1   Light is a wave, which is a form of electromagnetic radiation. In terms of waves, light can be described as a continuously repeating change, or oscillation, in electric and magnetic fields that can travel through space. Two characteristics of light are wavelength (often given in nanometers, nm), and frequency. The eye can see the light of wavelengths from about 400 nm (violet) to less than 800 nm (red).

7.2   The relationship among the different characteristics of light waves is $c = \nu\lambda$, where ν is the frequency, λ is the wavelength, and c is the speed of light.

7.3   The term "quantized" means the possible values of the energies of an atom are limited to only certain values. Planck was trying to explain the intensity of light of various frequencies emitted by a hot solid at different temperatures. The formula he arrived at was $E = nh\nu$, where E is energy, n is a whole number (n = 1, 2, 3, ...), h is Planck's constant, and ν is frequency.

7.4   Photoelectric effect is the term applied to the ejection of electrons from the surface of a metal or from other materials when light shines on it. Electrons are ejected only when the frequency (or energy) of light is larger than a certain minimum, or threshold, value that is constant for each metal. If a photon has a frequency equal to or greater than this minimum value, then it will eject one electron from the metal surface.

7.5   The wave-particle picture of light regards the wave and particle depictions of light as complementary views of the same physical entity. Neither view alone is a complete description of all the properties of light. The wave picture characterizes light only by wavelength and frequency. The particle picture characterizes light only as having an energy equal to $h\nu$.

7.6   The equation that relates the particle properties of light is $E = h\nu$. The symbol E is energy, h is Planck's constant, and ν is the frequency of the light.

7.7   According to physical theory at Rutherford's time, an electrically charged particle revolving around a center would continuously lose energy as electromagnetic radiation. As an electron in an atom lost energy, it would spiral into the nucleus (in about 10^{-10} s). Thus the stability of the atom could not be explained.

7.8   According to Bohr, an electron in an atom can have only specific energy values. An electron in an atom can change energy only by going from one energy level (of allowed energy) to another energy level (of allowed energy). An electron in a higher energy level can go to a lower energy level by emitting a photon of an energy equal to the difference in energy. However, when an electron is in its lowest energy level, no further changes in energy can occur. Thus the electron does not continuously radiate energy as thought at Rutherford's time. These features solve the difficulty alluded to in Question 7.7.

7.9   Emission of a photon occurs when an electron in a higher energy level undergoes a transition to a lower energy level. The energy lost is emitted as a photon.

7.10   Absorption of a photon occurs when a photon of a certain required energy is absorbed by a certain electron in an atom. The energy of the photon must be equal to the energy necessary to excite the electron of the atom from a lower energy level, usually the lowest, to a higher energy level. The photon's energy is converted into electronic energy.

7.11   The diffraction of an electron beam is evidence of electron waves. A practical example of diffraction is the operation of the electron microscope.

7.12 a. The principal quantum number can have an integer value between one and infinity.

     b. The angular momentum quantum number can have any integer value between zero and (n - 1).

     c. The magnetic quantum number can have any integer value between -l and +l.

     d. The spin quantum number can be either +1/2 or -1/2.

7.13   The notation is *3d*. This subshell contains five orbitals.

7.14   An *s* orbital has a spherical shape. A *p* orbital has two lobes positioned along a straight line through the nucleus at the center of the line (a dumbbell shape).

■ Solutions to Practice Problems

Note on significant figures: If the final answer to a solution needs to be rounded off, it is given first with one nonsignificant figure, and the last significant figure is underlined. The final answer is then rounded to the correct number of significant figures. In multiple-step problems, intermediate answers are given with at least one nonsignificant figure; however, only the final answer has been rounded off. Starting with Problem 7.21, the value 2.998×10^8 m/s will be used for the speed of light.

7.21   Solve $c = \lambda v$ for λ:

$$\lambda = \frac{c}{v} = \frac{2.998 \times 10^8 \text{ m/s}}{1.365 \times 10^6 \text{ /s}} = 219.\underline{6}3 = 219.6 \text{ m}$$

7.23   Solve $c = \lambda v$ for v. Recognize that 672 nm = 672×10^{-9} m, or 6.72×10^{-7} m.

$$v = \frac{c}{\lambda} = \frac{2.998 \times 10^8 \text{ m/s}}{6.72 \times 10^{-7} \text{ m}} = 4.4\underline{6}1 \times 10^{14} = 4.46 \times 10^{14} \text{ /s}$$

7.25   To do the calculation, divide one meter by the number of wavelengths in one meter to find the wavelength of this transition. Then use the speed of light (with nine digits for significant figures) to calculate the frequency:

$$\lambda = \frac{1 \text{ m}}{1,650,763.73} = 6.50780210\underline{6} \times 10^{-7} \text{ m}$$

$$v = \frac{c}{\lambda} = \frac{2.99792458 \times 10^8 \text{ m/s}}{6.507802106 \times 10^{-7} \text{ m}} = 4.94886516\underline{2} \times 10^{14}$$

$$= 4.94886516 \times 10^{14} \text{ /s}$$

7.27   Solve for E, using $E = hv$, and use four significant figures for h:

$$E = hv = (6.626 \times 10^{-34} \text{ J} \cdot \text{s}) \times (1.396 \times 10^6 \text{ /s}) = 9.24\underline{9}9 \times 10^{-28}$$

$$= 9.250 \times 10^{-28} \text{ J}$$

7.29 Recognize that 535 nm = 535 x 10^{-9} m = 5.35 x 10^{-7} m. Then calculate v and E.

$$\lambda = \frac{c}{v} = \frac{2.998 \times 10^8 \text{ m/s}}{5.35 \times 10^{-7} \text{ m}} = 5.6\underline{0}37 \times 10^{14} \text{ /s}$$

$$E = hv = (6.626 \times 10^{-34} \text{ J} \cdot \text{s}) \times (5.6037 \times 10^{14} \text{/s}) = 3.7\underline{1}3 \times 10^{-19} = 3.71 \times 10^{-19} \text{ J}$$

7.31 Solve the equation $E = -R_H/n^2$ for both E_5 and E_2; equate to hv, and solve for v.

$$E_5 = \frac{-R_H}{5^2} = \frac{-R_H}{25}; \qquad E_2 = \frac{-R_H}{2^2} = \frac{-R_H}{4}$$

$$\left[\frac{-R_H}{25}\right] - \left[\frac{-R_H}{4}\right] = \frac{21 R_H}{100} = hv$$

The frequency of the emitted radiation is:

$$v = \frac{21 R_H}{100 \, h} = \frac{21}{100} \times \frac{2.179 \times 10^{-18} \text{ J}}{6.626 \times 10^{-34} \text{ J} \cdot \text{s}} = 6.9\underline{0}5 \times 10^{14} = 6.91 \times 10^{14} \text{ /s}$$

7.33 This is the highest energy transition from the $n = 5$ level, so the electron must undergo a transition to the $n = 1$ level. Solve the Balmer equation using Bohr's approach:

$$E_5 = \frac{-R_H}{5^2} = \frac{-R_H}{25}; \qquad E_1 = \frac{-R_H}{1^2} = \frac{-R_H}{1}$$

$$\left[\frac{-R_H}{25}\right] - \left[\frac{-R_H}{1}\right] = \frac{24 R_H}{25} = hv$$

The frequency of the emitted radiation is:

$$v = \frac{24 R_H}{25 \, h} = \frac{24}{25} \times \frac{2.179 \times 10^{-18} \text{ J}}{6.626 \times 10^{-34} \text{ J} \cdot \text{s}} = 3.1\underline{5}7 \times 10^{15} \text{ /s}$$

The wavelength can now be calculated.

$$\lambda = \frac{c}{v} = \frac{2.998 \times 10^8 \text{ m/s}}{3.157 \times 10^{15} \text{ /s}} = 9.4\underline{9}6 \times 10^{-8} = 9.50 \times 10^{-8} \text{ m} \quad (95.0 \text{ nm})$$

7.35 Noting that 422.7 nm = 4.227 x 10^{-7} m, convert the 422.7 nm to frequency. Then convert the frequency to energy using $E = h\nu$.

$$\nu = \frac{c}{\lambda} = \frac{2.998 \times 10^8 \text{ m/s}}{4.227 \times 10^{-7} \text{ m}} = 7.09\underline{2}5 \times 10^{14} \text{ /s}$$

$$E = h\nu = (6.626 \times 10^{-34} \text{ J•s}) \times 7.09\underline{2}5 \times 10^{14} \text{ /s}) = 4.69\underline{9}4 \times 10^{-19}$$

$$= 4.699 \times 10^{-19} \text{ J}$$

7.37 The mass of a neutron = 1.67493 x 10^{-27} kg. Its speed or velocity, v, of 4.15 km/s equals 4.15 x 10^3 m/s. Substitute these parameters into the de Broglie relation, and solve for λ:

$$\lambda = \frac{h}{mv} = \frac{6.626 \times 10^{-34} \text{ kg} \cdot \text{m}^2/\text{s}}{1.67493 \times 10^{-27} \text{ kg} \times 4.15 \times 10^3 \text{ m/s}}$$

$$= 9.5\underline{3}2 \times 10^{-11} = 9.53 \times 10^{-11} \text{ m} \quad \text{or} \quad 95.3 \text{ pm}$$

7.39 The mass of an electron equals 9.10953 x 10^{-31} kg. The wavelength, λ, given as 20.0 pm, is equivalent to 2.00 x 10^{-11} m. Substitute these parameters into the de Broglie relation, and solve for the frequency, v.

$$v = \frac{h}{m\lambda} = \frac{6.626 \times 10^{-34} \text{ kg} \cdot \text{m}^2/\text{s}}{9.10953 \times 10^{-31} \text{ kg} \times 2.00 \times 10^{-11} \text{ m}} = 3.3\underline{6}3 \times 10^7$$

$$= 3.64 \times 10^7 \text{ m/s}$$

7.41 The possible values of l range from zero to $(n - 1)$, so l may be 0, 1, 2, or 3. The possible values of m_l range from $-l$ to $+l$, so m_l may be -3, -2, -1, 0, +1, +2, or +3.

7.43 a. 6d b. 5g c. 4f d. 6p

7.45 a. Not permissible; m_s may be only + 1/2 or -1/2.

b. Not permissible; l can only be as large as $(n - 1)$.

c. Not permissible; m_l may not exceed +2 in magnitude.

d. Not permissible; n may not be zero.

e. Not permissible; m_s may only be + 1/2 or -1/2.

■ Solutions to General Problems

7.47 Use $c = v\lambda$ to calculate frequency; then use $E = hv$ to calculate energy.

$$v = \frac{c}{\lambda} = \frac{2.998 \times 10^8 \text{ m/s}}{4.61 \times 10^{-7} \text{ m}} = 6.5\underline{0}3 \times 10^{14} = 6.50 \times 10^{14} \text{ /s}$$

$$E = hv = (6.626 \times 10^{-34} \text{ J}\cdot\text{s}) \times (6.503 \times 10^{14}\text{/s}) = 4.3\underline{0}9 \times 10^{-19}$$

$$= 4.31 \times 10^{-19} \text{ J}$$

7.49 Calculate the frequency corresponding to 4.34×10^{-19} J. Then convert that to wavelength.

$$v = \frac{E}{h} = \frac{4.34 \times 10^{-19} \text{ J}}{6.626 \times 10^{-34} \text{ J}\cdot\text{s}} = 6.5\underline{5}0 \times 10^{14} \text{ /s}$$

$$\lambda = \frac{c}{v} = \frac{2.998 \times 10^8 \text{ m/s}}{6.550 \times 10^{14} \text{ /s}} = 4.5\underline{7}7 \times 10^{-7} = 4.58 \times 10^{-7} \text{ m} = 458 \text{ nm (indigo)}$$

7.51 Solve for frequency using $E = hv$.

$$v = \frac{E}{h} = \frac{4.34 \times 10^{-19} \text{ J}}{6.626 \times 10^{-34} \text{ J}\cdot\text{s}} = 6.5\underline{4}9 \times 10^{14} = 6.55 \times 10^{14} \text{ /s}$$

7.53 This is a transition from the $n = 5$ level to the $n = 2$ level. Solve the Balmer equation using Bohr's approach.

$$E_5 = \frac{-R_H}{5^2} = \frac{-R_H}{25} ; \qquad E_2 = \frac{-R_H}{2^2} = \frac{-R_H}{4}$$

$$hv = \left[\frac{-R_H}{25}\right] - \left[\frac{-R_H}{4}\right] = \frac{21\,R_H}{100}$$

$$v = \frac{21\,R_H}{100\,h} = \frac{21}{100} \times \frac{2.179 \times 10^{-18} \text{ J}}{6.626 \times 10^{-34} \text{ J}\cdot\text{s}} = 6.9\underline{0}59 \times 10^{14} \text{ /s}$$

$$\lambda = \frac{c}{v} = \frac{2.998 \times 10^8 \text{ m/s}}{6.9059 \times 10^{14} \text{ /s}} = 4.3\underline{4}1 \times 10^{-7} = 4.34 \times 10^{-7} \text{ m (434 nm)}$$

7.55   Employ the Balmer formula using Z = 2 for the He⁺ ion.

$$E_3 = (2)^2 \; \frac{-R_H}{3^2} = (4) \; \frac{-R_H}{9} \; ; \qquad E_2 = (2)^2 \; \frac{-R_H}{2^2} = (4) \; \frac{-R_H}{4}$$

$$4 \left[\frac{-R_H}{9} \right] - 4 \left[\frac{-R_H}{4} \right] = 4 \times \frac{5\,R_H}{36} = h\nu$$

The frequency of the radiation is

$$\nu = \frac{4 \times 5\,R_H}{36\,h} = \frac{20}{36} \times \frac{2.179 \times 10^{-18}\ \text{J}}{6.626 \times 10^{-34}\ \text{J} \cdot \text{s}} = 1.8\underline{2}69 \times 10^{15}\ /\text{s}$$

$$\lambda = \frac{c}{\nu} = \frac{2.998 \times 10^8\ \text{m/s}}{1.8269 \times 10^{15}\ /\text{s}} = 1.6\underline{4}09 \times 10^{-7} = 1.64 \times 10^{-7}\ \text{m} \quad (164\ \text{nm; near UV})$$

7.57 a. Five   b. Seven   c. Three   d. One

7.59   The possible subshells for the $n = 5$ shell are $5s$, $5p$, $5d$, $5f$, and $5g$.

■ Solutions to Cumulative-Skills Problems

7.61   First use Avogadro's number to calculate the energy for one Cl_2 molecule.

$$\frac{239\ \text{kJ}}{1\ \text{mol}} \times \frac{1000\ \text{J}}{1\ \text{kJ}} \times \frac{1\ \text{mol}}{6.022 \times 10^{23}\ \text{molecules}} = 3.9\underline{6}87 \times 10^{-19}\ \text{J/molecule}$$

Then convert energy to frequency and finally to wavelength.

$$\nu = \frac{E}{h} = \frac{3.9687 \times 10^{-19}\ \text{J}}{6.626 \times 10^{-34}\ \text{J} \cdot \text{s}} = 5.9\underline{8}97 \times 10^{14}\ /\text{s}$$

$$\lambda = \frac{c}{\nu} = \frac{2.998 \times 10^8\ \text{m/s}}{5.9897 \times 10^{14}\ /\text{s}} = 5.0\underline{0}52 \times 10^{-7}\ \text{m} \quad (501\ \text{nm; visible region})$$

7.63 First calculate the energy needed to heat the 0.250 L of water from 20.0°C to 100.0°C.

$$0.250 \text{ L} \times \frac{1000 \text{ g}}{1 \text{ L}} \times \frac{4.184 \text{ J}}{(\text{g} \cdot °\text{C})} \times (100.0 \text{ °C} - 20.0°\text{C}) = 8.3\underline{6}8 \times 10^4 \text{ J}$$

Then calculate the frequency, the energy of one photon, and the number of photons.

$$\nu = \frac{c}{\lambda} = \frac{2.998 \times 10^8 \text{ m/s}}{0.125 \text{ m}} = 2.3\underline{9}8 \times 10^9 \text{ /s}$$

E of one photon $= h\nu = (6.626 \times 10^{-34} \text{ J} \cdot \text{s}) \times (2.398 \times 10^9 \text{/s}) = 1.5\underline{8}9 \times 10^{-24} \text{ J}$

No. photons $= h\nu = 8.368 \times 10^4 \text{ J} \times \dfrac{1 \text{ photon}}{1.589 \times 10^{-24} \text{ J}} = 5.2\underline{6}56 \times 10^{28}$

$$= 5.27 \times 10^{28} \text{ photons}$$

7.65 First write the following equality for the energy to remove one electron, $E_{removal}$:

$$E_{removal} = E_{425 \text{ nm}} - E_k \text{ of ejected photon}$$

Use $E = h\nu$ to calculate the energy of the photon. Then recall that E_k, the kinetic energy, is $1/2mv^2$. Use this to calculate E_k.

$$E_{425 \text{ nm}} = \frac{hc}{\lambda} = \frac{(6.626 \times 10^{-34} \text{ J} \cdot \text{s})(2.998 \times 10^8 \text{ m/s})}{4.25 \times 10^{-7} \text{ m}} = 4.6\underline{7}40 \times 10^{-19} \text{ J}$$

$E_k = 1/2mv^2 = 1/2 \times (9.1095 \times 10^{-31} \text{ kg}) \times (4.88 \times 10^5 \text{ m/s})^2 = 1.0\underline{8}46 \times 10^{-19} \text{ J}$

Subtract to find $E_{removal}$, and convert it to kJ/mol:

$$E_{removal} = 4.6740 \times 10^{-19} \text{ J} - (1.0847 \times 10^{-19} \text{ J}) = 3.5\underline{8}93 \times 10^{-19}$$

$$= 3.59 \times 10^{-19} \text{ J/electron}$$

$$E_{removal} = \frac{3.5893 \times 10^{-19} \text{ J}}{1 \text{ e}^-} \times \frac{6.022 \times 10^{23} \text{ e}^-}{1 \text{ mol}} \times \frac{1 \text{ kJ}}{1000 \text{ J}} = 2.1\underline{6}1 \times 10^2$$

$$= 2.16 \times 10^2 \text{ kJ/mol}$$

8. ELECTRON CONFIGURATIONS AND PERIODICITY

■ Answers to Review Questions

8.1   In the original Stern-Gerlach experiment, a beam of silver atoms is directed into the field of a specially designed magnet. (The same can be done with hydrogen atoms.) The beam of atoms is split into two by the magnetic field; half are bent toward one magnetic pole face and the other half toward the other magnetic pole face. This effect shows that the atoms themselves act as magnets with a positive or a negative component as indicated by the positive or negative spin quantum numbers.

8.2   In effect, the electron acts as though it were a sphere of spinning charge (Figure 8.3). Like any circulating electric charge, it creates a magnetic field with a spin axis that has more than one possible direction relative to a magnetic field. Electron spin is subject to a quantum restriction to one of two directions corresponding to the m_s quantum numbers +1/2 and -1/2.

8.3   According to the principles discussed in Section 7.5, the number of orbitals in the g subshell (l = 4) is given by 2l + 1 and is thus equal to nine. Because each orbital can hold a maximum of two electrons, the g subshell can hold a maximum of eighteen electrons.

8.4   The noble-gas core is an inner-shell configuration corresponding to one of the noble gases. The pseudo-noble-gas core is an inner-shell configuration corresponding to one of the noble gases together with $(n-1)d^{10}$ electrons. Like the noble-gas core electrons, the d^{10} electrons are not involved in chemical reactions. The valence electron is an electron (of an atom) located outside the noble-gas core or pseudo-noble-gas core. It is an electron primarily involved in chemical reactions.

8.5 The orbital diagram for the $1s^2 2s^2 2p^4$ ground state of oxygen is

1s   2s   2p

Another possible oxygen orbital diagram, but not a ground state, is

1s   2s   2p

8.6 A diamagnetic substance is a substance that is not attracted by a magnetic field or is very slightly repelled by such a field. This property generally indicates the substance has only paired electrons. A paramagnetic substance is a substance that is weakly attracted by a magnetic field. This property generally indicates the substance has one or more unpaired electrons. Ground-state oxygen has two unpaired $2p$ electrons and is therefore paramagnetic.

8.7 In Groups IA and IIA, the outer s subshell is being filled: s^1 for Group IA and s^2 for Group IIA. In Groups IIIA to VIIIA, the outer p subshell is being filled: p^1 for IIIA, p^2 for IVA, p^3 for VA, p^4 for VIA, p^5 for VIIA, and p^6 for VIIIA. In the transition elements, the $(n - 1)d$ subshell is being filled from d^1 to d^{10} electrons. In the lanthanides and actinides, the f subshell is being filled from f^1 to f^{14} electrons.

8.8 Mendeleev arranged the elements in increasing order of atomic weight, an arrangement that was later changed to atomic numbers. His periodic table was divided into rows (periods) and columns (groups). In his first attempt, he left spaces for what he believed to be undiscovered elements. In row five, under silicon and above tin in Group IV, he left a blank space. This Group IV element he called eka-silicon, and he predicted its properties from those of silicon and tin.

8.9 In a plot of atomic radii versus atomic number (Figure 8.16), the major trends that emerge are the following: (1) Within each period (horizontal row), the atomic radius tends to decrease with increasing atomic number or nuclear charge. The largest atom in a period is thus the Group 1A atom, and the smallest atom in a period is thus the noble-gas atom. (2) Within each group (vertical column), the atomic radius tends to increase with the period number.

In a plot of ionization energy versus atomic number (Figure 8.18), the major trends are (1) the ionization energy within a period increases with atomic number, and (2) the ionization energy within a group tends to decrease going down the group.

8.10 Group VIIA (halogens) is the main group with the most negative electron affinities. Configurations with filled subshells (ground states of the noble-gas elements) would form unstable negative ions when adding one electron per atom.

8.11   The Na^+ and Mg^{2+} ions are stable because they are isoelectronic with the noble gas neon. If Na^{2+} and Mg^{3+} ions were to exist, they would be very unstable because they would not be isoelectronic with any noble-gas structure and because of the energy needed to remove an electron from an inner shell.

8.12   The elements tend to increase in metallic character from right to left in any period. They also tend to increase in metallic character down any column (group) of elements.

8.13   A basic oxide is an oxide that reacts with acids. An example is calcium oxide, CaO. An acidic oxide is an oxide that reacts with bases. An example is carbon dioxide, CO_2.

8.14   Rubidium is the alkali metal atom with a $5s^1$ configuration.

8.15   Atomic number equals 117 (protons in last known element plus those needed to reach Group VIIA).

8.16   The following elements are in Groups IIIA to VIA:

Group IIIA	Group IVA	Group VA	Group VIA
B: metalloid	C: nonmetal	N: nonmetal	O: nonmetal
Al: metal	Si: metalloid	P: nonmetal	S: nonmetal
Ga: metal	Ge: metalloid	As: metalloid	Se: nonmetal
In: metal	Sn: metal	Sb: metalloid	Te: metalloid
Tl: metal	Pb: metal	Bi: metal	Po: metal

Yes, each column displays the expected increasing metallic character.

8.17   The oxides of the following elements are listed as either acidic, basic, amphoteric, or t.g.n.i. (text gives no information):

Group IIIA	Group IVA	Group VA	Group VIA
B: acidic	C: acidic	N: acidic	O: amphoteric (H_2O)
Al: amphoteric	Si: acidic	P: acidic	S: acidic
Ga: amphoteric	Ge: t.g.n.i.	As: t.g.n.i.	Se: acidic
In: basic	Sn: amphoteric	Sb: t.g.n.i.	Te: t.g.n.i.
Tl: basic	Pb: amphoteric	Bi: t.g.n.i.	Po: t.g.n.i.

8.18 $2Li(s) + 2H_2O(l) \rightarrow 2LiOH(aq) + H_2(g)$

8.19   Barium should be a soft, reactive metal. Barium should form the basic oxide, BaO. Barium metal, for example, would be expected to react with water according to the equation

$$Ba(s) + 2H_2O(l) \rightarrow Ba(OH)_2(aq) + H_2(g)$$

8.20 a. White phosphorus   b. Sulfur   c. Bromine   d. Sodium

■ Solutions to Practice Problems

Note on significant figures: If the final answer to a solution needs to be rounded off, it is given first with one nonsignificant figure, and the last significant figure is underlined. The final answer is then rounded to the correct number of significant figures. In multiple-step problems, intermediate answers are given with at least one nonsignificant figure; however, only the final answer has been rounded off.

8.27 a. Not allowed; the paired electrons in the $2p$ orbital should have opposite spins.

    b. Allowed; electron configuration is $1s^2 2s^2 2p^4$.

    c. Not allowed; the electrons in the $1s$ orbital must have opposite spins.

    d. Not allowed; the $2s$ orbital can hold only two electrons maximum, with opposite spins.

8.29 a. Impossible state; the $2p$ orbitals can hold no more than six electrons.

    b. Impossible state; the $3s$ orbital can hold no more than two electrons.

    c. Possible state.

    d. Possible state; however, the $3p$ and $4s$ orbitals should be filled before the $3d$ orbital.

8.31   Iodine ($Z = 53$): $1s^2 2s^2 2p^6 3s^2 3p^6 3d^{10} 4s^2 4p^6 4d^{10} 5s^2 5p^5$

8.33   Manganese ($Z = 25$): $1s^2 2s^2 2p^6 3s^2 3p^6 3d^5 4s^2$

8.35   Cadmium ($Z = 48$): $4d^{10} 5s^2$

8.37   Nickel ($Z = 28$): [Ar]

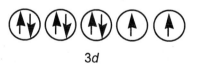

                  $3d$ $4s$

8.39   Rubidium ($Z = 37$): [Kr]

    $5s$

    All the subshells are filled in the krypton core; however, the $5s$ electron is unpaired, causing the ground state of the rubidium atom to be a paramagnetic substance.

8.41   Atomic radius increases going down a column (group), from S to Se, and increases going from right to left in a row, from Se to As. Thus the order by increasing atomic radius is S, Se, As.

8.43   Ionization energy increases going left to right in a row. Thus the order by increasing ionization energy is Na, Al, Cl, Ar.

8.45 a. In general, the electron affinity becomes more negative going from left to right within a period. Thus Br has a more negative electron affinity than As.

   b. In general, a nonmetal has a more negative electron affinity than a metal. Thus F has a more negative electron affinity than Li.

■ Solutions to General Problems

8.47   Polonium: $6s^2 6p^4$

8.49   The orbital diagram for arsenic is:

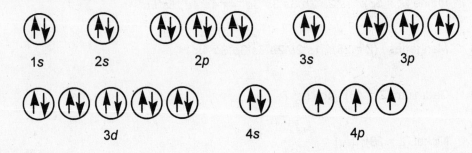

8.51   For eka-lead: [Rn] $5f^{14} 6d^{10} 7s^2 7p^2$. It is a metal; the oxide is eka-PbO or eka-PbO$_2$.

8.53   Niobium: [Kr]

4d   5s

8.55 a. Cl$_2$ b. Na   c. Sb   d. Ar

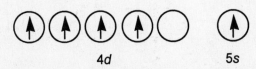

8.57   Element with Z = 23: $1s^2 2s^2 2p^6 3s^2 3p^6 3d^3 4s^2$. The element is in Group VB (three of the five valence electrons are d electrons) and in Period 4 (largest n is four). It is a d-block transition element.

■ Solutions to Cumulative-Skills Problems

8.59   The equation is:

$$Ba(s) + 2H_2O(l) \rightarrow Ba(OH)_2(aq) + H_2(g)$$

Using the equation, calculate the moles of H_2; then use the ideal gas law to convert to volume.

$$\text{mol } H_2 = 2.50 \text{ g Ba} \times \frac{1 \text{ mol Ba}}{137.33 \text{ g Ba}} \times \frac{1 \text{ mol } H_2}{1 \text{ mol Ba}} = 0.018\underline{2}04 \text{ mol}$$

$$V = \frac{nRT}{P} = \frac{(0.018204 \text{ mol})(0.082057 \text{ L} \cdot \text{atm/K} \cdot \text{mol})(294.2 \text{ K})}{(748/760) \text{ atm}}$$

$$= 0.446\underline{5}1 \text{ L} \quad (447 \text{ mL})$$

8.61   Radium is in Group IIA; hence the radium cation is Ra^{2+}, and its oxide is RaO. Use the atomic weights to calculate the percentage of Ra in RaO.

$$\text{Percent Ra} = \frac{226 \text{ amu Ra}}{226 \text{ amu Ra} + 16.00 \text{ amu O}} \times 100\% = 93.\underline{3}8 = 93.4\% \text{ Ra}$$

8.63   Convert 5.00 mg (0.00500 g) Na to moles of Na; then convert to energy using the first ionization energy of 496 kJ/mol Na.

$$\text{mol Na} = 0.00500 \text{ g Na} \times \frac{1 \text{ mol Na}}{22.99 \text{ g Na}} = 2.1\underline{7}4 \times 10^{-4} \text{ mol Na}$$

$$2.1\underline{7}4 \times 10^{-4} \text{ mol Na} \times \frac{496 \text{ kJ}}{1 \text{ mol Na}} = 0.10\underline{7}8 = 0.108 \text{ kJ} = 108 \text{ J}$$

8.65   Use the Bohr formula, where $n_f = \infty$ and $n_i = 1$.

$$\Delta E = -R_H \left[\frac{1}{\infty^2} - \frac{1}{1^2} \right] = -R_H[-1] = R_H = \frac{2.179 \times 10^{-18} \text{ J}}{1 \text{ H atom}}$$

$$\text{I.E.} = \frac{2.179 \times 10^{-18} \text{ J}}{1 \text{ H atom}} \times \frac{6.022 \times 10^{23} \text{ H atoms}}{1 \text{ mol H}} = \frac{1.31219 \times 10^6 \text{ J}}{1 \text{ mol H}}$$

$$= 1.312 \times 10^3 \text{ kJ/mol H}$$

9. IONIC AND COVALENT BONDING

■ Answers to Review Questions

9.1   As an Na atom approaches a Cl atom, the outer electron of the Na atom is transferred to the Cl atom. The result is an Na^+ and a Cl^- ion. Positively charged ions attract negatively charged ions, so, finally, the NaCl crystal consists of Na^+ ions surrounded by six Cl^- ions surrounded by six Na^+ ions.

9.2   The energy terms involved in the formation of an ionic solid from atoms are the ionization energy of the metal atom, the electron affinity of the nonmetal atom, and the energy of the attraction of the ions forming the ionic solid. The energy of the solid will be low if the ionization energy of the metal is low, the electron affinity of the nonmetal is high, and the energy of the attraction of the ions is large.

9.3   The lattice energy for potassium bromide is the change in energy that occurs when KBr(s) is separated into isolated $K^+(g)$ and $Br^-(g)$ ions in the gas phase.

$$KB(s) \rightarrow K^+(g) + Br^-(g)$$

9.4   A monatomic cation with a charge equal to the group number corresponds to the loss of all valence electrons. This loss of electrons would give a noble-gas configuration, which is especially stable. A monatomic anion with a charge equal to the group number minus eight would have a noble-gas configuration.

9.5   Most of the transition elements have configurations in which the outer s subshell is doubly occupied. These electrons will be lost first, and we might expect each to be lost with almost equal ease, resulting in +2 ions.

9.6   In going across a period, the cations decrease in radius. When we reach the anions, there is an abrupt increase in radius, and then the radii again decrease. Ionic radii increase going down any column of the periodic table.

9.7   An example is thionyl chloride, $SOCl_2$:

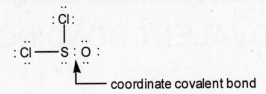

────── coordinate covalent bond

Note that the O atom has eight electrons around it; that is, it has two more electrons than the neutral atom. These two electrons must have come from the S atom. Thus this bond is a coordinate covalent bond.

9.8   In many atoms of the main-group elements, bonding uses an *s* orbital and the three *p* orbitals of the valence shell. These four orbitals are filled with eight electrons, thus accounting for the octet rule.

9.9   Electronegativity increases from left to right (with the exception of the noble gases) and decreases from top to bottom in the periodic table.

9.10   The absolute difference in the electronegativities of the two atoms in a bond gives a rough measure of the polarity of the bond.

9.11   Resonance is used to describe the electron structure of a molecule in which bonding electrons are delocalized. In a resonance description, the molecule is described in terms of two or more Lewis formulas. If we want to retain Lewis formulas, resonance is required because each Lewis formula assumes a bonding pair of electrons occupies the region between two atoms. We must imagine that the actual electron structure of the molecule is a composite of all resonance formulas.

9.12   Molecules having an odd number of electrons do not obey the octet rule. An example is nitrogen monoxide, NO. The other exceptions fall into two groups. In one group are molecules with an atom having fewer than eight valence electrons around it. An example is borane, BH_3. In the other group are molecules with an atom having more than eight valence electrons around it. An example is sulfur hexafluoride, SF_6.

9.13   As the bond order increases, the bond length decreases. For example, the average carbon-carbon single-bond length is 154 pm, whereas the carbon-carbon double-bond length is 134 pm, and the carbon-carbon triple-bond length is 120 pm.

9.14   Bond energy is the average enthalpy change for the breaking of a bond in a molecule. The enthalpy of a reaction for gaseous reactions can be determined by summing the bond energies of all the bonds that are broken and subtracting the sum of the bond energies of all the bonds that are formed.

■ Solutions to Practice Problems

9.23 a. As: $\qquad$ $1s^2 2s^2 2p^6 3s^2 3p^6 3d^{10} 4s^2 4p^3$ $\qquad$ $: \overset{\displaystyle\cdot}{As} \cdot$

b. As^{3+}: $\qquad$ $1s^2 2s^2 2p^6 3s^2 3p^6 3d^{10} 4s^2$ $\qquad$ $\left[: As \cdot \right]^{3+}$

c. Se: $\qquad$ $1s^2 2s^2 2p^6 3s^2 3p^6 3d^{10} 4s^2 4p^4$ $\qquad$ $: \overset{\displaystyle\cdot}{\underset{\displaystyle\cdot\cdot}{Se}} \cdot$

d. Se^{2-}: $\qquad$ $1s^2 2s^2 2p^6 3s^2 3p^6 3d^{10} 4s^2 4p^6$ $\qquad$ $\left[: \overset{\displaystyle\cdot\cdot}{\underset{\displaystyle\cdot\cdot}{Se}} : \right]^{2-}$

9.25 a. Bi: $\qquad$ $[Xe]4f^{14} 5d^{10} 6s^2 6p^3$

b. Bi^{3+}: $\qquad$ The three $6p$ electrons are lost from the valence shell.

$\qquad$ $[Xe]4f^{14} 5d^{10} 6s^2$

9.27 $\quad$ The +2 ion is formed by the loss of electrons from the $4s$ subshell.

$\qquad$ Ni^{2+}: $\quad$ $[Ar]3d^8$

The +3 ion is formed by the loss of electrons from the $4s$ and $3d$ subshells.

$\qquad$ Ni^{3+}: $\quad$ $[Ar]3d^7$

9.29 a. $Sr^{2+} < Sr$

The cation is smaller than the neutral atom because it has lost all its valence electrons; hence it has one less shell of electrons. The electron-electron repulsion is reduced, so the orbitals shrink because of the increased attraction of the electrons to the nucleus.

b. $Br < Br^-$

The anion is larger than the neutral atom because it has more electrons. The electron-electron repulsion is greater, so the valence orbitals expand to give a larger radius.

9.31 $\quad$ Smallest Na^+ ($Z = 11$), F^- ($Z = 9$), N^{3-} ($Z = 7$) Largest

These ions are isoelectronic. The atomic radius increases with the decreasing nuclear charge (Z).

9.33

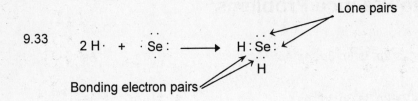

9.35 a. P, N, O

Electronegativity increases from left to right and bottom to top in the periodic table.

b. Na, Mg, Al

Electronegativity increases from left to right within a period.

c. Al, Si, C

Electronegativity increases from left to right and bottom to top in the periodic table.

9.37 $X_O - X_P = 3.5 - 2.1 = 1.4$

$X_{Cl} - X_C = 3.0 - 2.5 = 0.5$

$X_{Br} - X_{As} = 2.8 - 2.0 = 0.8$

The bonds arranged by increasing difference in electronegativity are C-Cl, As-Br, P-O.

9.39 a. Total number of valence electrons = 7 + 7 = 14. Br—Br is the skeleton. Distribute the remaining twelve electrons.

:Br — Br:

b. Total valence electrons = (2 x 1) + 6 = 8. The skeleton is H—Se—H. Distribute the remaining four electrons.

H——Se——H

(continued)

c. Total valence electrons = (3 x 7) + 5 = 26. The skeleton is

$$\begin{array}{c} \text{F} \\ | \\ \text{F——N——F} \end{array}$$

Distribute the remaining twenty electrons.

$$\begin{array}{c} :\ddot{\text{F}}: \\ | \\ :\ddot{\text{F}}——\text{N}——\ddot{\text{F}}: \end{array}$$

9.41 a. Total valence electrons = 7 + (2 x 6) + 1 = 20. The skeleton is O-Cl-O. Distribute the remaining electrons to achieve an octet.

$$\left[:\ddot{\text{O}}——\ddot{\text{Cl}}——\ddot{\text{O}}: \right]^{-}$$

b. Total valence electrons = 4 + (3 x 7) + 1 = 26. The skeleton is

$$\begin{array}{c} \text{Cl} \\ | \\ \text{Cl ——Sn—— Cl} \end{array}$$

Distribute the remaining twenty electrons so that each atom has an octet.

$$\left[\begin{array}{c} :\ddot{\text{Cl}}: \\ | \\ :\ddot{\text{Cl}}—— \text{Sn} ——\ddot{\text{Cl}}: \end{array} \right]^{-}$$

c. Total valence electrons = 6 + 5 -1 = 10. The compound requires a triple bond.

$$\left[:\text{N} \equiv \ddot{\text{O}}: \right]^{+}$$

9.43 a. There are two possible resonance structures for SO_2.

One pair of electrons is delocalized over the region of the two sulfur-oxygen bonds.

b. There are two possible resonance structures for HNO_3.

9.45 a. Total valence electrons = 8 + (2 x 7) = 22. The skeleton is F-Xe-F. Place six electrons around each fluorine atom to satisfy its octet.

There are three electron pairs remaining. Place them on the xenon atom.

b. Total valence electrons = 6 + (4 x 7) = 34. The skeleton is

Distribute twenty-four of the remaining twenty-six electrons on the fluorine atoms. The remaining pair of electrons is placed on the selenium atom.

(continued)

c. Total valence electrons = 6 + (6 x 7) = 48. The skeleton is

Distribute the remaining thirty-six electrons on the fluorine atoms.

d. Total valence electrons = 8 + (5 x 7) - 1 = 42. The skeleton is

Use thirty of the remaining thirty-two electrons on the fluorine atoms to complete their octets. The remaining two electrons form a lone pair on the xenon atom.

9.47 a. The total number of electrons in O_3 is 3 x 6 = 18. Assume a skeleton structure in which one oxygen atom is singly bonded to the other two oxygen atoms. This requires six electrons for the three single bonds, leaving twelve electrons to be used. It is impossible to fill the outer octets of all three oxygen atoms by writing three electron pairs around each, so a double bond must be written between the central oxygen and one of the other oxygen atoms. Then distribute the electron pairs to the oxygen atoms to satisfy the octet rule.

(continued)

As shown later, there are three bonds and six lone pairs of electrons, or eighteen electrons, in the structure. Thus all eighteen electrons are accounted for. One of the possible resonance structures is shown below; the other structure would have the double bond written between the left and central oxygen atoms.

$$ \overset{..}{\underset{..}{:O}} - \overset{..}{O} = \overset{..}{\underset{..}{O}} $$

Starting with the left oxygen, the formal charge of this oxygen is 6 - 1 - 6 = -1. The formal charge of just the central oxygen is 6 - 3 - 2 = +1. The formal charge of the right oxygen is 6 - 2 - 4 = 0. The sum of all three is zero.

b. The total number of electrons in CO is 4 + 6 = 10. Assume a skeleton structure in which the oxygen atom is singly bonded to carbon. This requires two electrons for the single bond and leaves eight electrons to be used. It is impossible to fill the outer octets of the carbon and oxygen by writing four electron pairs around each, so a triple bond must be written between the carbon and oxygen. Then distribute the electron pairs to both atoms to satisfy the octet rule. As shown below, there are one triple bond and two lone pairs of electrons, or ten electrons, in the structure. Thus all ten electrons are accounted for. The structure is

$$: C \equiv O : $$

The formal charge of the carbon is 4 - 3 - 2 = -1. The formal charge of the oxygen is 6 - 3 - 2 = +1. The sum of both is zero.

c. The total number of electrons in HNO_3 is 1 + 5 + 18 = 24. Assume a skeleton structure in which the nitrogen atom is singly bonded to two oxygen atoms and doubly bonded to one oxygen. This requires two electrons for the O—H single bond and leaves eight electrons to be used for the N bonds.

$$ H - \overset{..}{\underset{..}{O}} - \underset{|}{N} = \overset{..}{\underset{..}{O}} : $$
$$ \overset{..}{\underset{..}{:O}} : $$

The formal charge of the nitrogen is 5 - 4 - 0 = +1. The formal charge of the hydrogen is 1 - 1 - 0 = 0. The formal charge of the oxygen bonded to the hydrogen is 6 - 2 - 4 = 0. The formal charge of the other singly bonded oxygen is 6 - 1 - 6 = -1. The formal charge of the doubly bonded oxygen is 6 - 2 - 4 = 0.

9.49 a. $d_{C-H} = r_C + r_H$ = 77 pm + 37 pm = 114 pm

b. $d_{S-Cl} = r_S + r_{Cl}$ = 104 pm + 99 pm = 203 pm

c. $d_{Br-Cl} = r_{Br} + r_{Cl}$ = 114 pm + 99 pm = 213 pm

d. $d_{Si-O} = r_{Si} + r_O$ = 117 pm + 66 pm = 183 pm

9.51

In the reaction, a C=C double bond is converted to a C-C single bond. An H-Br bond is broken, and one C-H bond and one C-Br bond are formed.

$$\Delta H \cong BE(C{=}C) + BE(H\text{-}Br) - BE(C\text{-}C) - BE(C\text{-}H) - BE(C\text{-}Br)$$
$$= (602 + 362 - 346 - 411 - 285)\ kJ = -78\ kJ$$

■ Solutions to General Problems

9.53 a. Strontium is a metal, and oxygen is a nonmetal. The binary compound is likely to be ionic. Strontium, in Group IIA, forms Sr^{2+} ions; oxygen, from Group VIA, forms O^{2-} ions. The binary compound has the formula SrO and is named strontium oxide.

b. Carbon and bromine are both nonmetals; hence the binary compound is likely to be covalent. Carbon usually forms four bonds, and bromine usually forms one bond. The formula for the binary compound is CBr_4. It is called carbon tetrabromide.

c. Gallium is a metal, and fluorine is a nonmetal. The binary compound is likely to be ionic. Gallium is in Group IIIA and forms Ga^{3+} ions. Fluorine is in Group VIIA and forms F^- ions. The binary compound is GaF_3 and is named gallium(III) fluoride.

d. Nitrogen and bromine are both nonmetals; hence the binary compound is likely to be covalent. Nitrogen usually forms three bonds, and bromine usually forms one bond. The formula for the binary compound is NBr_3. It is called nitrogen tribromide.

9.55 Total valence electrons = 5 + (4 x 6) + 3 = 32. The skeleton is

(continued)

Distribute the remaining twenty-four electrons to complete the octets around the oxygen atoms.

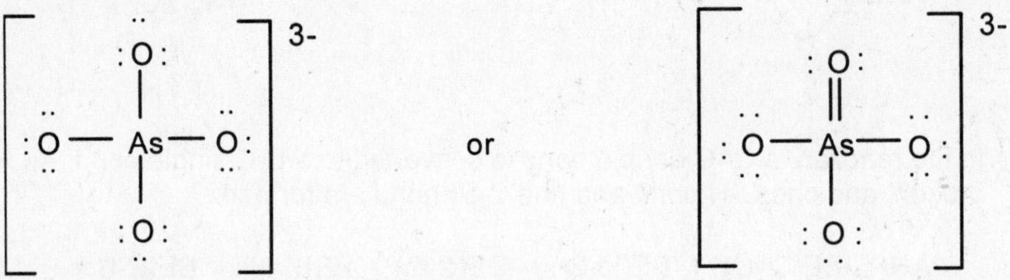

or

The formula for lead(II) arsenate is $Pb_3(AsO_4)_2$. The structure on the right has no formal charge.

9.57   Total valence electrons = 1 + 7 + (3 x 6) = 26. The skeleton is

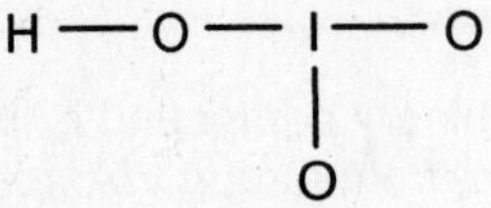

Distribute the remaining eighteen electrons to satisfy the octet rule. The structure on the right has no formal charge.

or

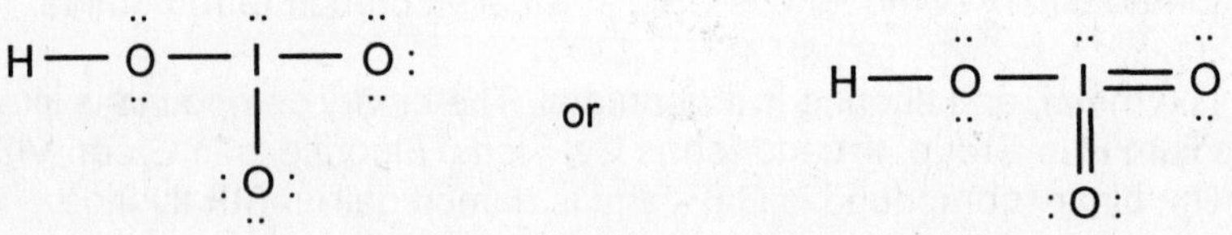

9.59   Total valence electrons = 5 + (2 x 1) + 1 = 8. The skeleton is H-N-H. Distribute the remaining four electrons to complete the octet of the nitrogen atom.

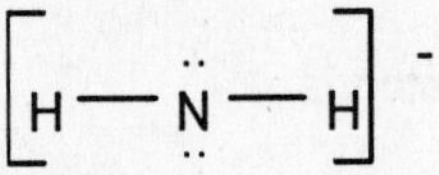

9.61 a. :Cl—Se—Cl: or   :Cl—Se—Cl: b. Se=C=Se

(with :O: above Se in first structure, and :O: double-bonded above Se in second structure)

c. [:Cl—Ga—Cl:] ⁻   with :Cl: above and :Cl: below Ga

d. [:C≡C:]²⁻

9.63 a. One possible electron-dot structure is

: O=Se—O:

Because the selenium-oxygen bonds are expected to be equivalent, the structure must be described in resonance terms.

: O=Se—O: ◄————► : O—Se=O:

One electron pair is delocalized over the selenium atom and the two oxygen atoms.

b. The possible electron-dot structures are

(four resonance structures of N—N with O atoms, connected by double-headed arrows)

At each end of the molecule, a pair of electrons is delocalized over the region of the nitrogen atom and the two oxygen atoms.

9.65   The possible electron-dot structures are

Because double bonds are shorter, the terminal N-O bonds are 118 pm, and the central N-O bonds are 136 pm.

9.67 ΔH = BE(N=N) + BE(F-F) - BE(N-N) - 2BE(N-F)

= (418 + 155 - 167 - 2 x 283) kJ = -1.60 x 10^2 kJ

■ Solutions to Cumulative-Skills Problems

Note on significant figures: The final answer to each cumulative-skills problem is given first with one nonsignificant figure (the rightmost significant figure is underlined) and then is rounded to the correct number of figures. Intermediate answers usually also have at least one nonsignificant figure. Atomic weights are rounded to two decimal places except for that of hydrogen.

9.69   The electronegativity differences and bond polarities are:

P- H	0.0	nonpolar
O- H	1.4	polar (acidic)

$H_3PO_3(aq)$ + 2NaOH(aq) $\rightarrow$ $Na_2HPO_3(aq)$ + 2H_2O(*l*)

$$\frac{mol}{L} = \frac{0.1250 \text{ mol NaOH}}{L \text{ NaOH}} \times 0.02250 \text{ L NaOH} \times \frac{1}{0.2000 \text{ L } H_3PO_3}$$

$$\times \frac{1 \text{ mol } H_3PO_3}{2 \text{ mol NaOH}} = 0.0070312 = 0.007031 \text{ M } H_3PO_3$$

9.71   After assuming a 100.0-g sample, convert to moles:

$$10.9 \text{ g Mg} \times \frac{1 \text{ mol Mg}}{24.3 \text{ g Mg}} = 0.448\underline{5} \text{ mol Mg}$$

$$31.8 \text{ g Cl} \times \frac{1 \text{ mol Cl}}{35.453 \text{ g Cl}} = 0.896\underline{9}6 \text{ mol Cl}$$

$$57.3 \text{ g O} \times \frac{1 \text{ mol O}}{16.00 \text{ g O}} = 3.5\underline{8}1 \text{ mol O}$$

Divide by 0.4485:

Mg: $\dfrac{0.4485}{0.4485} = 1$; Cl: $\dfrac{0.89696}{0.4485} = 2.00$; O: $\dfrac{3.581}{0.4485} = 7.98$

The simplest formula is $MgCl_2O_8$. However, since $Cl_2O_8{}^{2-}$ is not a well known ion, write the simplest formula as $Mg(ClO_4)_2$, magnesium perchlorate. The Lewis formulas are Mg^{2+} and

9.73   After assuming a 100.0-g sample, convert to moles:

$$25.0 \text{ g C} \times \frac{1 \text{ mol C}}{12.01 \text{ g C}} = 2.0\underline{8}1 \text{ mol C}$$

$$2.1 \text{ g H} \times \frac{1 \text{ mol H}}{1.008 \text{ g H}} = 2.\underline{0}8 \text{ mol H}$$

$$39.6 \text{ g F} \times \frac{1 \text{ mol F}}{18.99 \text{ g F}} = 2.0\underline{8}5 \text{ mol F}$$

$$33.3 \text{ g O} \times \frac{1 \text{ mol O}}{16.00 \text{ g O}} = 2.0\underline{8}1 \text{ mol O}$$

(continued)

The simplest formula is CHOF. Because the molecular mass of 48.0 divided by the formula mass of 48.0 is one, the molecular formula is also CHOF. The Lewis formula is

9.75 $HCN(g)$ → $H(g)$ + $C(g)$ + $N(g)$

(135   218   715.0   473) ΔH_f (kJ/mol)

ΔH_{rxn} = 1271 kJ/mol

BE(C≡N) = ΔH - BE(C-H) = [1271 - 411] kJ/mol

= 860 kJ/mol (Table 9.5 has 887 kJ/mol.)

10. MOLECULAR GEOMETRY AND CHEMICAL BONDING THEORY

■ Answers to Review Questions

10.1 The VSEPR model is used to predict the geometry of molecules. The electron pairs around an atom are assumed to arrange themselves to reduce electron repulsion. The molecular geometry is determined by the positions of the bonding electron pairs.

10.2 The arrangements are linear, trigonal planar, tetrahedral, trigonal bipyramidal, and octahedral.

10.3 A lone pair is "larger" than a bonding pair; therefore, it will occupy an equatorial position where it encounters less repulsion than if it were in an axial position.

10.4 The bonds could be polar, but if they are arranged symmetrically the molecule will be nonpolar. The bond dipoles will cancel.

10.5 Certain orbitals, such as p orbitals and hybrid orbitals, have lobes in given directions. Bonding to these orbitals is directional; that is, the bonding is in preferred directions. This explains why the bonding gives a particular molecular geometry.

10.6 A sigma bond has a cylindrical shape about the bond axis. A pi bond has a distribution of electrons above and below the bond axis.

10.7 In ethylene, C₂H₄, the changes on a given carbon atom may be described as follows:

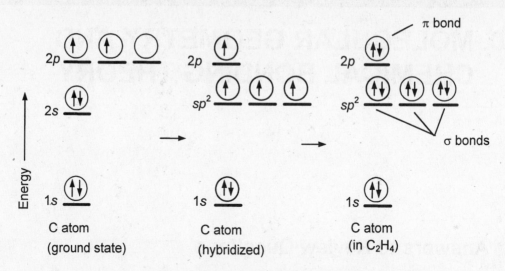

An *sp²* hybrid orbital on one carbon atom overlaps a similar hybrid orbital on the other carbon atom to form a σ bond. The remaining hybrid orbitals on the two carbon atoms overlap 1s orbitals from the hydrogen atoms to form four C-H bonds. The unhybridized 2p orbital on one carbon atom overlaps the unhybridized 2p orbital on the other carbon atom to form a π bond. The σ and π bonds together constitute a double bond.

10.8 Both of the unhybridized 2p orbitals, one from each carbon atom, are perpendicular to their CH₂ planes. When these orbitals overlap each other, they fix both planes in the same plane. The two ends of the molecule cannot twist around without breaking the π bond, which requires considerable energy. Therefore, it is possible to have stable molecules with the following structures:

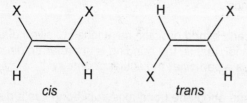

Because these have the same molecular formulas, they are isomers. In this case, they are called *cis-trans* isomers, or geometrical isomers.

10.9 In a bonding orbital, the probability of finding electrons between the two nuclei is high. For this reason, the energy of the bonding orbital is lower than that of the separate atomic orbitals. In an antibonding orbital, the probability of finding electrons between the two nuclei is low. For this reason, the energy of the antibonding orbital is higher than that of the separate atomic orbitals.

10.10 The factors determining the strength of interaction of two atomic orbitals are (1) the energy difference between the interacting orbitals and (2) the magnitude of their overlap.

10.11 When two 2s orbitals overlap, they interact to form a bonding orbital, σ_{2s}, and an antibonding orbital, $\sigma_{2s}{}^{*}$. The bonding orbital is at lower energy than the antibonding orbital.

10.12 When two 2p orbitals overlap, they interact to form a bonding sigma orbital, σ_{2p}, and an antibonding sigma orbital, $\sigma_{2p}{}^{*}$. Two bonding pi orbitals, π_{2p} (each having the same energy), and two antibonding pi orbitals, $\pi_{2p}{}^{*}$ (each having the same energy), are also formed.

■ Solutions to Practice Problems

10.17 The number of electron pairs, lone pairs and geometry are for the central atom.

	Electron-Dot Structure	Number of Electron Pairs	Number of Lone Pairs	Geometry
a.		4	0	Tetrahedral
b.		4	2	Bent
c.		3	0	Trigonal planar
d.		4	1	Trigonal pyramidal

10.19 The number of electron pairs, lone pairs and geometry are for the central atom.

Electron-Dot Structure	Number of Electron Pairs	Number of Lone Pairs	Geometry
a.	4	1	Trigonal pyramidal
b.	4	0	Tetrahedral
c.	2	0	Linear
d.	4	1	Trigonal pyramidal

10.21 The number of electron pairs, lone pairs and geometry are for the central atom.

Electron-Dot Structure	Number of Electron Pairs	Number of Lone Pairs	Geometry
a.	5	0	Trigonal bipyramidal
b.	5	2	T-shaped

(continued)

c.

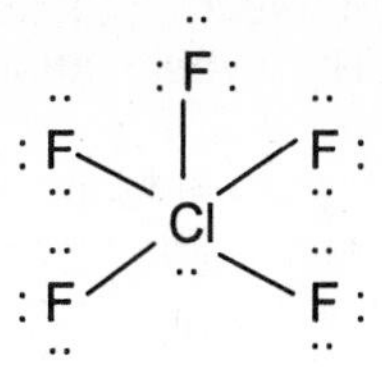

$$: \overset{..}{\underset{..}{F}} :$$

Square pyramidal   6   1

d.

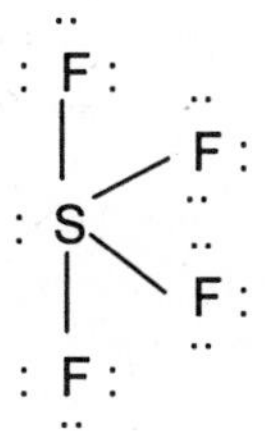

Seesaw   5   1

10.23 a. Trigonal pyramidal and T-shaped are both possible. Trigonal planar would have a dipole moment of zero. VSEPR theory predicts a trigonal pyramidal geometry for AsF_3.

   b. Bent. Linear would have a dipole moment of zero.

10.25 a. SF_2 is an angular model. The sulfur atom is sp^3 hybridized.

   b. ClO_3^- is a trigonal pyramidal molecule. The chlorine atom is sp^3 hybridized.

10.27 a. The Lewis structure is

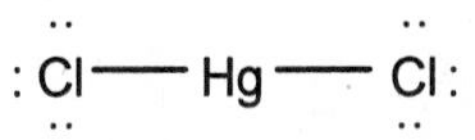

$$: Cl — Hg — Cl :$$

The presence of two single bonds and no lone pairs suggests sp hybridization. Thus, an Hg atom with the configuration $[Xe]4f^{14}5d^{10}6s^2$ is promoted to $[Xe]4f^{14}5d^{10}6s^16p^1$, then hybridized. An Hg–Cl bond is formed by overlapping an Hg hybrid orbital with a 3_l orbital of Cl.

   b. The Lewis structure is

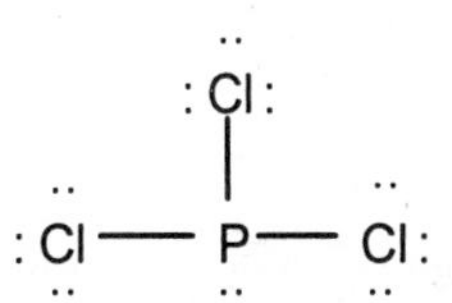

$$: Cl — P — Cl :$$
$$: Cl :$$

The presence of three single bonds and one lone pair suggests sp^3 hybridization of the P atom. Three hybrid orbitals each overlap a $3p$ orbital of a Cl atom to form a P–Cl bond. The fourth hybrid orbital contains the lone pair.

10.29 a. Xenon has eight valence electrons. Each F atom donates one electron to give a total of ten electrons, or five electron pairs, around the Xe atom. The hybridization is sp^3d.

b. Bromine has seven valence electrons. Each F atom donates one electron to give a total of twelve electrons, or six electron pairs, around the Br atom. The hybridization is sp^3d^2.

c. Phosphorus has five valence electrons. The Cl atoms each donate one electron to give a total of ten electrons, or five electron pairs, around the P atom. The hybridization is sp^3d.

d. Chlorine has seven valence electrons, to which may be added one electron from each F atom minus one electron for the charge on the ion. This gives a total of ten electrons, or five electron pairs, around chlorine. The hybridization is sp^3d.

10.31 The P atom in PCl_6^- has six single bonds around it and no lone pairs. This suggests sp^3d^2 hybridization. Each bond in this ion is a σ bond formed by overlap of an sp^3d^2 hybrid orbital on P with a $3p$ orbital on Cl.

10.33 Each of the N atoms has a lone pair of electrons and is bonded to two atoms. The N atoms are sp^2 hybridized. The two possible arrangements of the O atoms relative to one another are shown below. Because the π bond between the N atoms must be broken to interconvert these two forms, it is to be expected that the hyponitrite ion will exhibit *cis-trans* isomerism.

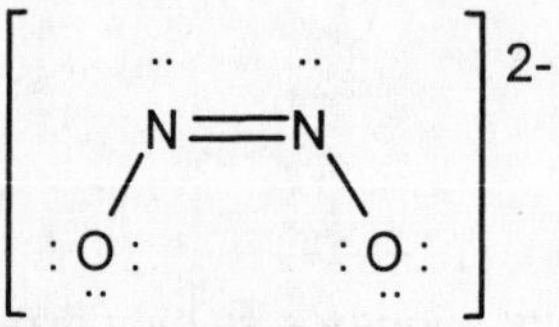

cis 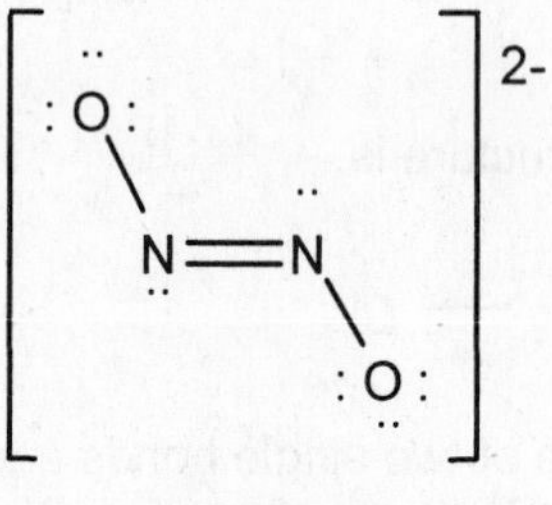   *trans*

10.35 a. Total electrons = 2 x 5 = 10.

The electron configuration is $KK(\sigma_{2s})^2(\sigma_{2s}{}^*)^2(\pi_{2p})^2$.

Bond order = $1/2 (n_b - n_a)$ = $1/2 (6 - 4)$ = 1

The B_2 molecule is stable. It is paramagnetic because the two electrons in the π_{2p} subshell occupy separate orbitals.

(continued)

b. Total electrons $= (2 \times 5) - 1 = 9$.

The electron configuration is $KK(\sigma_{2s})^2(\sigma_{2s}*)^2(\pi_{2p})^1$.

Bond order $= 1/2 \ (n_b - n_a) = 1/2 \ (5 - 4) = 1/2$

The B_2^+ molecule should be stable and is paramagnetic because there is one unpaired electron in the π_{2p} subshell.

c. Total electrons $= (2 \times 8) + 1 = 17$.

The electron configuration is $KK(\sigma_{2s})^2(\sigma_{2s}*)^2(\pi_{2p})^4(\sigma_{2p})^2(\pi_{2p}*)^3$.

Bond order $= 1/2 \ (n_b - n_a) = 1/2 \ (10 - 7) = 3/2$

The O_2^- molecule should be stable and is paramagnetic because there is one unpaired electron in the $\pi_{2p}*$ subshell.

■ Solutions to General Problems

10.37   The number of electron pairs, lone pairs and geometry are for the central atom.

	Electron- Dot Structure	Number of Electron Pairs	Number of Lone Pairs	Geometry
a.	$:Cl\!-\!Ge\!-\!Cl:$	3	1	Bent
b.	(structure with $:O:$ double bonded to C, with H and H)	3	0	Trigonal planar
c.	$\left[:I\!-\!I\!-\!I:\right]^{-}$	5	3	Linear
d.	(structure with P central, six Cl)	6	0	Octahedral

10.39 a.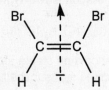

C_a and C_b: Three electron pairs around each. They are sp^2 hybridized.

C_c: Four electron pairs around it. It is sp^3 hybridized.

b. :N≡C—C≡N:

Both C atoms are bonded to two other atoms and have no lone pairs of electrons. They are sp hybridized.

10.41

Br—C=C—Br (with H, H)   *cis*   has a net dipole

Br—C=C—Br (with H, H)

H—C=C—Br (with Br, H)   *trans*   has no net dipole. The two C-Br bond dipoles cancel and the two C-H bond dipoles cancel.

10.43 Total electrons = 2 + 1 - 1 = 2.

The electron configuration is $(\sigma_{1s})^2$.

Bond order = $1/2 \, (n_b - n_a)$ = 1/2 (2) = 1

The HeH$^+$ ion is expected to be stable.

10.45 Total electrons = (2 x 6) + 2 = 14.

The electron configuration is $KK(\sigma_{2s})^2(\sigma_{2s}{}^*)^2(\pi_{2p})^4(\sigma_{2p})^2$.

Bond order = $1/2 \, (n_b - n_a)$ = 1/2 (10 - 4) = 3

10.47　The molecular orbital configuration of O_2 is $KK(\sigma_{2s})^2(\sigma_{2s}{}^*)^2(\pi_{2p})^4(\sigma_{2p})^2(\pi_{2p}{}^*)^2$. $O_2{}^+$ has one electron less than O_2. The difference is in the number of electrons in the $\pi_{2p}{}^*$ antibonding orbital. This means the bond order is larger for $O_2{}^+$ than for O_2.

O_2: Bond order $= 1/2\ (n_b - n_a) = 1/2\ (10 - 6) = 2$

$O_2{}^+$: Bond order $= 1/2\ (n_b - n_a) = 1/2\ (10 - 5) = 5/2$

It is expected that the species with the higher bond order, $O_2{}^+$, has the shorter bond length. In $O_2{}^-$, there is one more electron than in O_2. This additional electron occupies a $\pi_{2p}{}^*$ orbital. Increasing the number of electrons in antibonding orbitals decreases the bond order; hence, $O_2{}^-$ should have a longer bond length than O_2.

■ Solutions to Cumulative-Skills Problems

Note on significant figures: If the final answer to a solution needs to be rounded off, it is given first with one nonsignificant figure, and the last significant figure is underlined. The final answer is then rounded to the correct number of significant figures. In multiple-step problems, intermediate answers are given with at least one nonsignificant figure; however, only the final answer has been rounded off.

10.49　After assuming a 100.0 g sample, convert to moles:

$$60.4\ \text{g Xe} \times \frac{1\ \text{mol Xe}}{131.29\ \text{g Xe}} = 0.46\underline{0}05\ \text{mol Xe}$$

$$22.1\ \text{g O} \times \frac{1\ \text{mol O}}{16.00\ \text{g O}} = 1.3\underline{8}1\ \text{mol O}$$

$$17.5\ \text{g F} \times \frac{1\ \text{mol F}}{18.99\ \text{g F}} = 0.92\underline{1}5\ \text{mol F}$$

Divide by 0.460 :

$$\text{Xe:}\ \frac{0.460}{0.460} = 1;\ \ \text{O:}\ \frac{1.381}{0.460} = 3.00;\ \ \text{F:}\ \frac{0.9215}{0.460} = 2.00$$

(continued)

The simplest formula is XeO_3F_2. This is also the molecular formula. The Lewis formula is

Number of electron pairs = 5, number of lone pairs = 0; hence, the geometry is trigonal bipyramidal. Because xenon has five single bonds, it will require five orbitals to describe the bonding. This suggests sp^3d hybridization.

10.51 N_2:   Triple bond; bond length = 110 pm. Geometry is linear; sp-hybrid orbitals are needed for one lone pair and one σ bond.

N_2F_2:   Double bond; bond length = 122 pm. Geometry is trigonal planar; sp^2-hybrid orbitals are needed for one lone pair and two σ bonds.

N_2H_4:   Single bond; bond length = 145 pm. Geometry is tetrahedral; sp^3-hybrid orbitals are needed for one lone pair and three σ bonds.

10.53 HNO_3 resonance formulas:

The geometry around the nitrogen is trigonal planar; therefore, the hybridization is sp^2.

Formation reaction: $H_2(g) + 3O_2(g) + N_2(g) \rightarrow 2HNO_3(g)$

$2 \times \Delta H_f° = [BE(H\text{-}H) + 3BE(O_2) + BE(N_2)] - [2BE(H\text{-}O) + 4BE(N\text{-}O) + 2BE(N\text{=}O)]$

$= [(432 + 3 \times 494 + 942) - (2 \times 459 + 4 \times 201 + 2 \times 607) \text{ kJ/2 mol}]$

$= -80 \text{ kJ/2 mol} = -40 \text{ kJ/mol}$

Resonance energy $= -40 \text{ kJ} - (-135 \text{ kJ}) = 95 \text{ kJ}$

11. STATES OF MATTER; LIQUIDS AND SOLIDS

■ Answers to Review Questions

11.1   The six different phase transitions, with examples in parentheses, are melting (snow melting), sublimation (dry ice subliming directly to carbon dioxide gas), freezing (water freezing), vaporization (water evaporating), condensation (dew forming on the ground), and gas-solid condensation or deposition (frost forming on the ground).

11.2   The vapor pressure of a liquid is the partial pressure of the vapor over the liquid, measured at equilibrium. In molecular terms, vapor pressure involves molecules of a liquid vaporizing from the liquid phase, colliding with any surface above the liquid, and exerting pressure on it. The equilibrium is a dynamic one because molecules of the liquid are continually leaving the liquid phase and returning to it from the vapor phase.

11.3   Steam at 100°C will melt more ice than the same weight of water at 100°C because it contains much more energy in the form of its heat of vaporization. It will transfer this energy to the ice and condense in doing so. The condensed steam and the water will both transfer heat to the ice as the temperature then drops.

11.4   The heat of fusion is smaller than the heat of vaporization because melting requires only enough energy for molecules to escape from their sites in the crystal lattice, leaving other molecular attractions intact. In vaporization, sufficient energy must be added to break almost all molecular attractions and also to do the work of pushing back the atmosphere.

11.5   Evaporation leads to cooling of a liquid because the gaseous molecules require heat to evaporate; as they leave the other liquid molecules, they remove the heat energy required to vaporize them. This leaves less energy in the liquid whose temperature then drops.

11.6   As the temperature increases for a liquid and its vapor in the closed vessel, the two, which are separated by a meniscus, gradually become identical. The meniscus first becomes fuzzy and then disappears altogether as the temperature reaches the critical temperature. Above this temperature, only the vapor exists.

11.7   A permanent gas can be liquefied only by lowering the temperature below its critical temperature while compressing the gas.

11.8   The pressure in the cylinder of nitrogen at room temperature (above its critical temperature of -147°C) decreases continuously as gas is released because the number of molecules in the vapor phase, which governs the pressure, decreases continuously. The pressure in the cylinder of propane at room temperature (below its critical temperature) is constant because liquid propane and gaseous propane exist at equilibrium in the cylinder. The pressure will remain constant at the vapor pressure of propane until only gaseous propane remains. At that point, the pressure will decrease until all of the propane is gone.

11.9   Hydrogen bonding is a weak to moderate attractive force that exists between a hydrogen atom covalently bonded to a very electronegative atom, X (N, O, or F), and a lone pair of electrons on another small, electronegative atom, Y. (X and Y may be the same or different elements.) Hydrogen bonding in water involves a hydrogen atom of one water molecule bonding to a lone pair of electrons on the oxygen atom of another water molecule.

11.10   Molecular substances have relatively low melting points because the forces broken by melting are weak intermolecular attractions in the solid state, not strong bonding attractions.

11.11   In a face-centered cubic cell, there are atoms at the center of each face of the unit cell in addition to those at the corners.

11.12   The structure of thallium(I) iodide is a simple cubic lattice for both the metal ions and the anions. Thus, the structure consists of two interpenetrating cubic lattices of cation and anion.

11.13   The coordination number of Cs^+ in CsCl is eight; the coordination number of Na^+ in NaCl is six; and the coordination number of Zn^{2+} in ZnS is four.

11.14   Starting with the edge length of a cubic crystal, we can calculate the volume of a unit cell by cubing the edge length. Then, knowing the density of the crystalline solid, we can calculate the mass of the atoms in the unit cell. Then, the mass of the atoms in the unit cell is divided by the number of atoms in the unit cell to give the mass of one atom. Dividing the mass of one mole of the crystal by the mass of one atom yields a value for Avogadro's number.

■ Solutions to Practice Problems

Note on significant figures: If the final answer to a solution needs to be rounded off, it is given first with one nonsignificant figure, and the last significant figure is underlined. The final answer is then rounded to the correct number of significant figures. In multiple-step problems, intermediate answers are given with at least one nonsignificant figure; however, only the final answer has been rounded off.

11.21 a. Vaporization

    b. Freezing of eggs and sublimation of ice

    c. Condensation

    d. Gas-solid condensation, deposition

11.23 Dropping a line from the intersection of a 250 mmHg line with the diethyl ether curve in Figure 11.6 intersects the temperature axis at about 5°C.

11.25 The heat absorbed per 2.25 g of isopropyl alcohol, C_3H_8O, is

$$2.25 \text{ g } C_3H_8O \times \frac{1 \text{ mol } C_3H_8O}{60.09 \text{ g } C_3H_8O} \times \frac{42.1 \text{ kJ}}{1 \text{ mol } C_3H_8O} = 1.5\underline{7}6 = 1.58 \text{ kJ}$$

11.27 Calculate how much heat is released by cooling 64.3 g of H_2O from 55°C to 15°C.

$$\text{Heat rel'd} = (64.3 \text{ g})(15°C - 55°C)\left(\frac{4.18 \text{ J}}{1 \text{ g} \cdot °C}\right)$$

$$= -1.\underline{0}7509 \times 10^4 \text{ J} = -1\underline{0}.7509 \text{ kJ}$$

The heat released is used first to melt the ice and then to warm the liquid from 0°C to 15°C. Let the mass of ice equal y grams. Then, for fusion and for warming, we have

$$\text{Fusion: (y g } H_2O) \times \frac{1 \text{ mol } H_2O}{18.02 \text{ g } H_2O} \times \frac{6.01 \text{ kJ}}{1 \text{ mol } H_2O} = 0.33\underline{3}5 \text{ y kJ}$$

$$\text{Warming: (y g } H_2O)(15°C - 0°C)\left(\frac{4.18 \text{ J}}{1 \text{ g} \cdot °C}\right) = 6\underline{2}.70 \text{ y J} \quad (0.06\underline{2}70 \text{ y kJ})$$

(continued)

Because the total heat required for melting and warming must equal the heat released by cooling, equate the two, and solve for y.

$$10.7509 \text{ kJ} = 0.3335y \text{ kJ} + 0.0627y \text{ kJ} = y(0.3335 + 0.0627) \text{ kJ}$$

$$y = 10.7509 \text{ kJ} \div 0.3962 \text{ kJ} = 27.13 \text{ (grams)}$$

Thus, 27 g of ice were added.

11.29 At the normal boiling point, the vapor pressure of a liquid is 760.0 mmHg. Use the Clausius-Clapeyron equation to find P_2 when P_1 = 760.0 mmHg, T_1 = 334.85 K (61.7°C), and T_2 = 309.35 K (36.2°C). Also use ΔH_{vap} = 31.4 x 10^3 J/mol.

$$\ln \frac{P_2}{P_1} = \frac{\Delta H_{vap}}{R}\left(\frac{1}{T_1} - \frac{1}{T_2}\right)$$

$$\ln \frac{P_2}{760 \text{ mmHg}} = \frac{31.4 \times 10^3 \text{ J/mol}}{8.31 \text{ J/K} \cdot \text{mol}}\left(\frac{1}{334.85 \text{ K}} - \frac{1}{309.35 \text{ K}}\right) = -0.93018$$

Taking antilogs of both sides gives

$$\frac{P_2}{760 \text{ mmHg}} = e^{-0.93018} = 0.3944$$

$$P_2 = 0.3944 \times 760 \text{ mmHg} = 299.8 = 3.00 \times 10^2 \text{ mmHg (300. mmHg)}$$

11.31 From the Clausius-Clapeyron equation,

$$\Delta H_{vap} = R\left(\frac{T_2 T_1}{T_2 - T_1}\right)\left[\ln \frac{P_2}{P_1}\right]$$

$$= [8.31 \text{ J/(K}\cdot\text{mol)}]\left[\frac{(553.2 \text{ K})(524.2 \text{ K})}{(553.2 - 524.2) \text{ K}}\right]\left[\ln \frac{760.0 \text{ mmHg}}{400.0 \text{ mmHg}}\right]$$

$$= 5.3336 \times 10^4 \text{ J/mol} = 53.3 \text{ kJ/mol}$$

11.33 a. At point A, the substance will be a gas.

b. The substance will be a gas.

c. The substance will be a liquid.

d. no

11.35 Liquefied at 25°C: SO_2 and C_2H_2. To liquefy CH_4, lower its temperature below -82°C, and then compress it. To liquefy CO, lower its temperature below -140°C, and then compress it.

11.37 a. London forces

    b. London and dipole-dipole forces, H-bonding

    c. London and dipole-dipole forces

11.39 CCl_4 has the lowest vapor pressure because it has the largest molecular weight and the greatest London forces even though $HCCl_3$ and H_3CCl have dipole-dipole interactions.

11.41 The order of decreasing vapor pressure is FCH_2CH_2F > FCH_2CH_2OH > $HOCH_2CH_2OH$; there is no hydrogen bonding in the first molecule for the weakest interaction and the highest vapor pressure; the second molecule can hydrogen-bond at only one end; and the third molecule can hydrogen-bond at both ends for the strongest interaction and lowest vapor pressure.

11.43 The order is CH_4 < C_2H_6 < CH_3OH < CH_2OHCH_2OH. The weakest forces are the London forces in CH_4 and C_2H_6, which increase with molecular weight. The next strongest interaction is in CH_3OH, which can hydrogen-bond at only one end of the molecule. The strongest interaction is in the last molecule, which can hydrogen-bond at both ends.

11.45 a. Metallic

    b. Covalent network (like diamond)

    c. Molecular

11.47 The order is $(C_2H_5)_2O$ < C_4H_9OH < KCl < CaO. Melting points increase in the order of attraction between molecules or ions in the solid state. Hydrogen bonding in C_4H_9OH causes it to melt at a higher temperature than $(C_2H_5)_2O$. Both KCl and CaO are ionic solids with much stronger attraction than the organic molecules. In CaO, the higher charges cause the lattice energy to be higher than in KCl.

11.49 a. LiCl   **b.** SiC   **c.** CHI_3 **d.** Co

11.51 In a face-centered cubic lattice with one atom at each lattice point, there are atoms at the corners and in the center of each face of the unit cell. Each corner is shared by eight unit cells, and there are eight corners per unit cell. Each face is shared by two unit cells, and there are six faces per unit cell. Therefore, there are four atoms per unit cell.

11.53 There are four Ni atoms in the face-centered cubic structure, so the mass of one cell is

$$4 \text{ Ni atoms} \times \frac{1 \text{ mol Ni}}{6.022 \times 10^{23} \text{ Ni atoms}} \times \frac{58.7 \text{ g Ni}}{1 \text{ mol Ni}} = 3.8\underline{9}9 \times 10^{-22} \text{ g}$$

$$\text{Cell volume} = \frac{3.899 \times 10^{-22} \text{ g}}{8.91 \text{ g/cm}^3} = 4.3\underline{7}6 \times 10^{-23} \text{ cm}^3$$

All edges are the same length in a cubic cell, so the edge length, l, is

$$l = \sqrt[3]{V} = \sqrt[3]{4.376 \times 10^{-23} \text{ cm}^3} = 3.5\underline{2}3 \times 10^{-8}$$

$$= 3.52 \times 10^{-8} \text{ cm} \quad (352 \text{ pm})$$

11.55 Calculate the volume from the edge length of 407.9 pm (4.079×10^{-8} cm), and then use it to calculate the mass of the unit cell:

$$\text{Cell volume} = (4.079 \times 10^{-8} \text{ cm}^3 = 6.7869 \times 10^{-23} \text{ cm}^3$$

$$\text{Cell mass} = (19.3 \text{ g/cm}^3)(6.7869 \times 20^{-23} \text{ cm}^3) = 1.3\underline{0}98 \times 10^{-21} \text{ g}$$

Calculate the mass of one gold atom:

$$1 \text{ Au atom} \times \frac{1 \text{ mol Au}}{6.022 \times 10^{23} \text{ Au atoms}} \times \frac{196.97 \text{ g Au}}{1 \text{ mol Au}} = 3.27\underline{0}8 \times 10^{-22} \text{ g}$$

$$\frac{1.3098 \times 10^{-21} \text{ g}}{1 \text{ unit cell}} \times \frac{1 \text{ Au atom}}{3.2708 \times 10^{-22} \text{ g Au}} = \frac{4.0\underline{0}4 \text{ Au atoms}}{\text{unit cell}}$$

Because there are four atoms per unit cell, it is a face-centered cubic.

11.57 Calculate the volume from the edge (316.5 pm = 3.165×10^{-8} cm). Use it to calculate the mass:

Cell volume = $(3.165 \times 10^{-8} \text{ cm})^3$ = $3.1705 \times 10^{-23} \text{ cm}^3$

For a body-centered cubic lattice, there are two atoms per cell, so their mass is

$$2 \text{ W atoms } \times \frac{1 \text{ mol W}}{6.022 \times 10^{23} \text{ W atoms}} \times \frac{183.8 \text{ g W}}{1 \text{ mol W}} = 6.1043 \times 10^{-22} \text{ g W}$$

$$\text{Density} = \frac{6.1043 \times 10^{-22} \text{ g W}}{3.1705 \times 10^{-23} \text{ cm}^3} = 19.253 = 19.25 \text{ g/cm}^3$$

■ Solutions to General Problems

11.59 From Table 5.5, the vapor pressures are 18.7 mmHg at 21°C and 12.8 mmHg at 15°C. If the moisture did not begin to condense until the air had been cooled to 15°C, then the partial pressure of water in the air at 21°C must have been 12.8 mmHg. The relative humidity is

$$\text{Percent relative humidity} = \frac{12.8 \text{ mmHg}}{18.7 \text{ mmHg}} \times 100\% = 68.44 = 68.4 \text{ percent}$$

11.61 After labeling the problem data as below, use the Clausius-Clapeyron equation to obtain ΔH_{vap}, which can then be used to calculate the boiling point.

At T_1 = 325.05 K, P_1 = 100.0 mmHg; at T_2 = 362.65 K, P_2 = 400.0 mmHg

$$\ln \frac{400.0 \text{ mmHg}}{100.0 \text{ mmHg}} = \frac{\Delta H_{vap}}{8.31 \text{ J/(K} \cdot \text{mol)}} \left(\frac{1}{325.05 \text{ K}} - \frac{1}{362.65 \text{ K}} \right)$$

$1.3862 = \Delta H_{vap} (3.838 \times 10^{-5} \text{ mol/J})$

$\Delta H_{vap} = 36.11 \times 10^3 \text{ J/mol} (36.11 \text{ kJ/mol})$

Now, use this value of ΔH_{vap} and the following data to calculate the boiling point:

At T_1 = 325.1 K, P_1 = 100.0 mmHg; at T_2 (boiling pt.), P_2 = 760 mmHg

(continued)

$$\ln\frac{760.0 \text{ mmHg}}{100.0 \text{ mmHg}} = \frac{36.11 \times 10^3 \text{ J/mol}}{8.314 \text{ J/(K} \cdot \text{mol)}} \left(\frac{1}{325.1 \text{ K}} - \frac{1}{T_2} \right)$$

$$2.0281 = 4.344 \times 10^3 \text{ K} \left(\frac{1}{325.05 \text{ K}} - \frac{1}{T_2} \right)$$

$$\frac{1}{T_2} = \frac{1}{325.05 \text{ K}} - \frac{2.0281}{4.344 \times 10^3 \text{ K}} = 2.6090 \times 10^{-3}/\text{K}$$

$$T_2 = 383.2 = 383 \text{ K } (110°\text{C})$$

11.63 a. As this gas is compressed at 20°C, it will condense into a liquid because 20°C is above the triple point but below the critical point.

 b. As this gas is compressed at -70°C, it will condense directly to the solid phase because the temperature of -70°C is below the triple point.

 c. As this gas is compressed at 40°C, it will not condense because 40°C is above the critical point.

11.65 In propanol, hydrogen bonding exists between the hydrogen of the OH group and the lone pair of electrons of oxygen of the OH group of an adjacent propanol molecule. For two adjacent propanol molecules, the hydrogen bond may be represented as follows:

 $C_3H_7-O-H\bullet\bullet\bullet O(H)C_3H_7$

11.67 Aluminum (Group IIIA) forms a metallic solid. Silicon (Group IVA) forms a covalent network solid. Phosphorus (Group VA) forms a molecular solid. Sulfur (Group VIA) forms a molecular (amorphous) solid.

11.69 a. Lower: KCl. The lattice energy should be lower for ions with a lower charge. A lower lattice energy implies a lower melting point.

 b. Lower: CCl_4. Both are molecular solids, so the compound with the lower molecular weight should have weaker London forces and, therefore, the lower melting point.

 c. Lower: Zn. Melting points for Group IIB metals are lower than for metals near the middle of the transition-metal series.

 d. Lower: C_2H_5Cl. Ethyl chloride cannot hydrogen bond, but acetic acid can. The compound with the weaker intermolecular forces has the lower melting point.

11.71   The face-centered cubic structure means one atom is at each lattice point. All edges are the same length in such a structure, so the volume is

$$\text{Volume} = l^3 = (3.839 \times 10^{-8} \text{ cm})^3 = 5.6579 \times 10^{-23} \text{ cm}^3$$

$$\text{Mass of unit cell} = dV = (22.42 \text{ g/cm}^3)(5.6579 \times 10^{-23} \text{ cm}^3) = 1.2685 \times 10^{-21} \text{ g}$$

There are four atoms in a face-centered cubic cell, so

$$\text{Mass of one Ir atom} = \text{mass of unit cell} \div 4 = (1.2685 \times 10^{-21} \text{ g}) \div 4$$

$$= 3.1712 \times 10^{-22} \text{ g}$$

$$\text{Molar mass of Ir} = (3.1712 \times 10^{-22} \text{ g/Ir atom}) \times (6.022 \times 10^{23} \text{ Ir atoms/mol})$$

$$= 190.96 = 191.0 \text{ g/mol} \quad (\text{The atomic weight} = 191.0 \text{ amu.})$$

11.73   From Problem 11.53, the cell edge length (l) is 352.3 pm. There are four nickel atom radii along the diagonal of a unit-cell face. Because the diagonal square $= l^2 + l^2$ (Pythagorean theorem),

$$4r = \sqrt{2\,l^2} = \sqrt{2}\,l, \text{ or } r = \frac{\sqrt{2}}{4}(352.3 \text{ pm}) = 124.5 = 125 \text{ pm}$$

11.75 a.   Diamond and silicon carbide are giant molecules with strong covalent bonds among all the atoms. Graphite is a layered structure, and the forces holding the layers together are weak dispersion forces.

   b.   Silicon dioxide is a giant molecule with an infinite array of O–S–O bonds. Each silicon is bonded to four oxygen atoms in a covalent network solid. Carbon dioxide is a discrete, nonpolar molecule.

11.77 a. CO_2 consists of discrete non-polar molecules, held together in the solid by weak dispersion forces. SiO_2 is a giant molecule with all the atoms held together by strong covalent bonds.

   b.   HF(l) has extensive hydrogen bonding among the molecules. HCl(l) boils much lower because it doesn't have H-bonding.

   c. SiF_4 is a larger molecule (it has more electrons than CF_4), so it has stronger dispersion forces and a higher boiling point than CF_4. Both molecules are tetrahadrally symmetrical and, therefore, non-polar.

■ Solutions to Cumulative-Skills Problems

11.79 Calculate the moles of HCN in 10.0 mL of the solution (density = 0.687 g HCN/mL HCN):

$$10.0 \text{ mL HCN} \times \frac{0.687 \text{ g HCN}}{1 \text{ mL HCN}} \times \frac{1 \text{ mol HCN}}{27.03 \text{ g HCN}} = 0.254\underline{1} \text{ mol HCN}$$

$$0.2541 \text{ mol HCN}(l) \quad \rightarrow \quad 0.2541 \text{ mol HCN}(g)$$

$$(\Delta H_f^\circ = 105 \text{ kJ/mol}) \qquad (\Delta H_f^\circ = 135 \text{ kJ/mol})$$

$$\Delta H^\circ = 0.2541 \text{ mol} \times [135 \text{ kJ/mol} - 105 \text{ kJ/mol}] = 7.\underline{6}23 = 7.6 \text{ kJ}$$

11.81 First convert the mass to moles; then multiply by the standard heat of formation to obtain the heat absorbed in vaporizing this mass:

$$12.5 \text{ g P}_4 \times \frac{1 \text{ mol P}_4}{123.88 \text{ g P}_4} = 0.100\underline{9} \text{ mol P}_4$$

$$0.1009 \text{ mol P}_4 \times \frac{95.4 \text{ J}}{°C \bullet \text{ mol P}_4} \times (44.1°C - 25.0°C)$$

$$= 18\underline{3}.86 \text{ J} = 0.18\underline{3}86 \text{ kJ}$$

$$2.63 \text{ kJ/mol P}_4 \times 0.1009 \text{ mol P}_4 = 0.26\underline{5}37 \text{ kJ}$$

$$\text{Total heat} = 0.26537 \text{ kJ} + 0.18386 \text{ kJ} = 0.44\underline{9}23 = 0.449 \text{ kJ} = 449 \text{ J}$$

11.83 Use the ideal gas law to calculate the total number of moles of monomer and dimer:

$$n = \frac{PV}{RT} = \frac{(436/760)\text{atm} \times 1.000 \text{ L}}{[0.082057 \text{ L} \bullet \text{atm}/(K \bullet \text{mol})] \times 373.75 \text{ K}}$$

$$= 0.018\underline{7}057 \text{ mol monomer} + \text{dimer}$$

$$(0.018\underline{7}057 \text{ mol monomer} + \text{dimer}) \times \frac{0.630 \text{ mol dimer}}{1 \text{ mol dimer} + \text{monomer}}$$

$$= 0.011\underline{7}8 \text{ mol dimer}$$

$$0.0187057 \text{ mol both} - 0.01178 \text{ mol dimer} = 0.006\underline{9}2 \text{ mol monomer}$$

(continued)

Mass dimer = 0.01178 mol dimer x $\dfrac{120.1 \text{ g dimer}}{1 \text{ mol dimer}}$ = 1.414 g dimer

Mass monomer = 0.00692 mol monomer x $\dfrac{60.05 \text{ g monomer}}{1 \text{ mol monomer}}$

= 0.4155 g monomer

Density = $\dfrac{1.414 \text{ g} + 0.4155 \text{ g}}{1.000 \text{ L}}$ = 1.829 = 1.83 g/L vapor

12. SOLUTIONS

■ Answers to Review Questions

12.1 An example of a gaseous solution is air, in which nitrogen (78 percent) acts as a solvent for a gas such as oxygen (21 percent). Recall that the solvent is the component present in greater amount. An example of a liquid solution containing a gas is any carbonated beverage in which water acts as the solvent for carbon dioxide gas. Ethanol in water is an example of a liquid-liquid solution. An example of a solid solution is any gold-silver alloy.

12.2 The two factors that explain differences in solubilities are (1) the natural tendency of substances to mix together, or the natural tendency of substances to become disordered, and (2) the relative forces of attraction between solute species (or solvent species) compared to that between the solute and solvent species. The strongest interactions are always achieved.

12.3 A sodium chloride crystal dissolves in water because of two factors. The positive Na^+ ion is strongly attracted to the oxygen (negative end of the water dipole) and dissolves as $Na^+(aq)$. The negative Cl^- ion is strongly attracted to the hydrogens (positive end of the water dipole) and dissolves as $Cl^-(aq)$.

12.4 When the temperature (energy of a solution) is increased, the solubility of an ionic compound usually increases. A number of salts are exceptions to this rule, particularly a number of calcium salts such as calcium acetate, calcium sulfate, and calcium hydroxide (although the solubilities of calcium bromide, calcium chloride, calcium fluoride, and calcium iodide all increase with temperature).

12.5 Calcium chloride is an example of a salt that releases heat when it dissolves (exothermic heat of solution). Ammonium nitrate is an example of a salt that absorbs heat when it dissolves (endothermic heat of solution).

12.6 As the temperature of the solution is increased by heating, the concentration of the dissolved gas would decrease.

12.7 According to Le Châtelier's principle, a gas is more soluble in a liquid at higher pressures because, when the gas dissolves in the liquid, the system decreases in volume, tending to decrease the applied pressure. However, when a solid dissolves in a liquid, there is very little volume change. Thus pressure has very little effect on the solubility of a solid in a liquid.

12.8 The four ways to express the concentration of a solute in a solution are (1) molarity, which is moles per liter; (2) mass percentage of solute, which is the percentage by mass of solute contained in a given mass of solution; (3) molality, which is the moles of solute per kilogram of solvent; and (4) mole fraction, which is the moles of the component substance divided by the total moles of solution.

12.9 The boiling point of the solution is higher because the nonvolatile solute lowers the vapor pressure of the solvent. Thus the temperature must be increased to a value greater than the boiling point of the pure solvent to achieve a vapor pressure equal to atmospheric pressure.

12.10 One application is the use of ethylene glycol in automobile radiators as antifreeze; the glycol-water mixture usually has a freezing point well below the average low temperature during the winter. A second application is spreading sodium chloride on icy roads in the winter to melt the ice. The ice usually melts because, at equilibrium, a concentrated solution of NaCl usually freezes at a temperature below that of the water.

12.11 If a pressure greater than the osmotic pressure of the ocean water is applied, the natural osmotic flow can be reversed. Then the water solvent flows from the ocean water through a membrane to a more dilute solution or to pure water, leaving behind the salt and other ionic compounds from the ocean in a more concentrated solution.

12.12 The polar -OH group on glycerol allows it to interact (hydrogen bond) with the polar water molecules, which means it is like water and, therefore, will dissolve in water. Benzene is a nonpolar molecule, which indicates it is not "like" water and, therefore, will not dissolve in water.

■ Solutions to Practice Problems

Note on significant figures: If the final answer to a solution needs to be rounded off, it is given first with one nonsignificant figure, and the last significant figure is underlined. The final answer is then rounded to the correct number of significant figures. In multiple-step problems, intermediate answers are given with at least one nonsignificant figure; however, only the final answer has been rounded off. Atomic weights are rounded to two decimal places, except for that of hydrogen.

12.19 An example of a liquid solution prepared by dissolving a gas in a liquid is household ammonia, which consists of ammonia (NH_3) gas dissolved in water.

12.21 The order of increasing solubility is H_2O < CH_2OHCH_2OH < $C_{10}H_{22}$. The solubility in nonpolar hexane increases with the decreasing polarity of the solute.

12.23 The Al^{3+} ion has both a greater charge and a smaller ionic radius than Mg^{2+}, so Al^{3+} should have a greater energy of hydration.

12.25 Using Henry's law, let S_1 = the solubility at 1.00 atm (P_1), and S_2 = the solubility at 5.50 atm (P_2).

$$S_2 = \frac{P_2 S_1}{P_1} = \frac{(5.50\ \text{atm})(0.161\ \text{g/100 mL})}{1.00\ \text{atm}} = 0.88\underline{5}5$$

$$= 0.886\ \text{g/100 mL}$$

12.27 First, calculate the mass of KI in the solution; then calculate the mass of water needed.

$$\text{Mass KI} = 72.5\ \text{g} \times \frac{5.00\ \text{g KI}}{100\ \text{g solution}} = 3.6\underline{2}50 = 3.63\ \text{g KI}$$

Mass H_2O = 72.5 g soln - 3.6250 g KI = 68.\underline{8}75 = 68.9 g H_2O

Dissolve 3.63 g KI in 68.9 g of water.

12.29 Convert mass of vanillin ($C_8H_8O_3$, molar mass 152.14 g/mol) to moles, convert mg of ether to kg, and divide for molality.

$$0.0391\ \text{g vanillin} \times \frac{1\ \text{mol vanillin}}{152.14\ \text{g vanillin}} = 2.5\underline{7}0 \times 10^{-4}\ \text{mol vanillin}$$

$$168.5\ \text{mg ether} \times 1\ \text{kg/10}^6\ \text{mg} = 168.5 \times 10^{-6}\ \text{kg ether}$$

$$\text{Molality} = \frac{2.570 \times 10^{-4}\ \text{mol vanillin}}{168.5 \times 10^{-6}\ \text{kg ether}} = 1.5\underline{2}52 = 1.53\ m\ \text{vanillin}$$

12.31 Convert mass of fructose ($C_6H_{12}O_6$, molar mass 180.16 g/mol) to moles, and then multiply by one kg H_2O per 0.125 mol fructose (the reciprocal of molality).

$$1.75\ \text{g fructose} \times \frac{1\ \text{mol fructose}}{180.16\ \text{g fructose}} \times \frac{1\ \text{kg}\ H_2O}{0.125\ \text{mol fructose}}$$

$$= 0.077\underline{7}0\ \text{kg}\quad (77.7\ \text{g}\ H_2O)$$

12.33 In the solution, for every 0.550 mol of NaClO there is 1.00 kg, or 1.00×10^3 g, H_2O, so

$$1.00 \times 10^3 \text{ g } H_2O \ \times \ \frac{1 \text{ mol } H_2O}{18.02 \text{ g } H_2O} \ = \ 55.\underline{4}9 \text{ mol } H_2O$$

Total mol $= 55.\underline{4}9$ mol H_2O + 0.550 mol NaClO $= 56.\underline{0}4$ mol

$$\text{Mol fraction NaClO} \ = \ \frac{\text{mol NaClO}}{\text{total mol}} \ = \ \frac{0.550 \text{ mol}}{56.04 \text{ mol}} \ = \ 0.0098\underline{1}3 \ = \ 0.00981$$

12.35 In 1.000 L of vinegar, there is 0.763 mole of acetic acid. The total mass of the 1.000 L solution is 1.004 kg. Start by calculating the mass of acetic acid (AA) in the solution.

$$0.763 \text{ mol AA } \times \ \frac{60.05 \text{ g AA}}{1 \text{ mol AA}} \ = \ 45.\underline{8}2 \text{ g AA} \ (0.045\underline{8}2 \text{ kg AA})$$

The mass of water may be found by difference:

Mass H_2O = 1.004 kg soln - 0.04582 kg AA = 0.95\underline{8}2 kg H_2O

$$\text{Molality} \ = \ \frac{0.763 \text{ mol AA}}{0.9582 \text{ kg } H_2O} \ = \ 0.79\underline{6}2 \ = \ 0.796 \ m \text{ AA}$$

12.37 To find the mole fraction of sucrose, first find the amounts of both sucrose (suc.) and water:

$$20.2 \text{ g sucrose } \times \ \frac{1 \text{ mol sucrose}}{342.30 \text{ g sucrose}} \ = \ 0.059\underline{0}1 \text{ mol sucrose}$$

$$70.1 \text{ g } H_2O \ \times \ \frac{1 \text{ mol } H_2O}{18.02 \text{ g } H_2O} \ = \ 3.8\underline{9}0 \text{ mol } H_2O$$

$$X_{\text{sucrose}} \ = \ \frac{0.05901 \text{ mol sucrose}}{(3.890 \ + \ 0.05901) \text{ mol}} \ = \ 0.014\underline{9}4$$

From Raoult's law, the vapor pressure (P) and lowering (ΔP) are

$$P \ = \ P^{\circ}_{H_2O} X_{H_2O} \ = \ P^{\circ}_{H_2O}(1 - X_{\text{suc.}}) \ = \ (42.2 \text{ mmHg})(1 \ - \ 0.01494)$$

$$= \ 41.\underline{5}69 \ = \ 41.6 \text{ mmHg}$$

$$\Delta P \ = \ P^{\circ}_{H_2O} X_{\text{suc.}} \ = \ (42.2 \text{ mmHg})(0.01494) \ = \ 0.63\underline{0}6 \ = \ 0.631 \text{ mmHg}$$

12.39 Calculate ΔT_f, the freezing-point depression, and, using K_f = 1.858°C/m (Table 12.2), the molality, c_m.

$$\Delta T_f = 0.000°C - (-0.104°C) = 0.104°C$$

$$c_m = \frac{\Delta T_f}{K_f} = \frac{0.104°C}{1.858°C/m} = 0.05597 = 0.560 \; m$$

12.41 Calculate ΔT_f, the freezing-point depression, and, using K_f = 1.858°C/m (Table 12.2), the molality, c_m.

$$\Delta T_f = 26.84°C - 25.70°C = 1.14°C$$

$$c_m = \frac{\Delta T_f}{K_f} = \frac{1.14°C}{8.00°C/m} = 0.1425 \; m$$

Find the moles of solute by rearranging the definition of molality:

$$Mol = c_m \times kg \; solvent = \frac{0.1425 \; mol}{1 \; kg \; solvent} \times 103 \times 10^{-6} \; kg \; solvent$$

$$= 1.467 \times 10^{-5} \; mol$$

$$Molar \; mass = \frac{2.39 \times 10^{-3} \; g}{1.467 \times 10^{-5} \; mol} = 162.9 = 163 \; g/mol$$

The molecular weight is 163 amu.

12.43 Use the equation for osmotic pressure (π) to solve for the molarity of the solution.

$$M = \frac{\pi}{RT} = \frac{1.47 \; mmHg \times \left(\dfrac{1 \; atm}{760 \; mmHg} \right)}{(0.0821 \; L \bullet atm/K \bullet mol)(21 + 273)K} = 8.013 \times 10^{-5} \; mol/L$$

Now find the number of moles in 106 mL (0.106 L) using the molarity.

$$0.106 \; L \times \frac{8.013 \times 10^{-5} \; mol}{1 \; L} = 8.494 \times 10^{-6} \; mol$$

$$Molar \; mass = \frac{0.582 \; g}{8.494 \times 10^{-6} \; mol} = 6.851 \times 10^4 = 6.85 \times 10^4 \; g/mol$$

The molecular weight is 6.85 x 10^4 amu.

12.45 Begin by calculating the molarity of $Cr(NH_3)_5Cl_3$.

$$1.40 \times 10^{-2} \text{ g } Cr(NH_3)_5Cl_3 \times \frac{1 \text{ mol } Cr(NH_3)_5Cl_3}{243.5 \text{ g } Cr(NH_3)_5Cl_3}$$

$$= 5.7\underline{4}9 \times 10^{-5} \text{ mol } Cr(NH_3)_5Cl_3$$

$$\text{Molarity} = \frac{5.749 \times 10^{-5} \text{ mol } Cr(NH_3)_5Cl_3}{0.0250 \text{ L}} = 0.0022\underline{9}9 \text{ M}$$

Now find the hypothetical osmotic pressure, assuming $Cr(NH_3)_5Cl_3$ does not ionize:

$$\pi = MRT = (2.30 \times 10^{-3} \text{ M}) \times \frac{0.0821 \text{ L} \cdot \text{atm}}{K \cdot \text{mol}} \times 298 \text{ K} \times \frac{760 \text{ mmHg}}{1 \text{ atm}}$$

$$= 42.\underline{7}7 \text{ mmHg}$$

The measured osmotic pressure is greater than the hypothetical osmotic pressure. The number of ions formed per formula unit equals the ratio of the measured pressure to the hypothetical pressure:

$$i = \frac{119 \text{ mmHg}}{42.77 \text{ mmHg}} = 2.7\underline{8}2 \cong 3 \text{ ions/formula unit}$$

■ Solutions to General Problems

12.47 Assume a volume of 1.000 L whose mass is then 1.036 kg. Use the percent composition given to find the mass of each of the components of the solution.

$$1.036 \text{ kg soln} \times \frac{12.00 \text{ kg } NH_4Cl}{100.00 \text{ kg soln}} = 0.124\underline{3}2 \text{ kg } NH_4Cl$$

Mass of H_2O = 1.036 kg soln - 0.12432 kg NH_4Cl = 0.91\underline{1}6 kg H_2O

Convert mass of NH_4Cl and water to moles:

$$124.32 \text{ g } NH_4Cl \times \frac{1 \text{ mol } NH_4Cl}{53.49 \text{ g } NH_4Cl} = 2.3\underline{2}4 \text{ mol } NH_4Cl$$

$$911.6 \text{ g } H_2O \times \frac{1 \text{ mol } H_2O}{18.015 \text{ g } H_2O} = 50.\underline{6}0 \text{ mol } H_2O$$

(continued)

$$\text{Molarity} = \frac{\text{mol NH}_4\text{Cl}}{\text{L solution}} = \frac{2.324 \text{ mol}}{1.00 \text{ L}} = 2.3\underline{2}4 = 2.32 \text{ } M$$

$$\text{Molality} = \frac{\text{mol NH}_4\text{Cl}}{\text{kg H}_2\text{O}} = \frac{2.324 \text{ mol}}{0.9116 \text{ kg H}_2\text{O}} = 2.5\underline{4}9 = 2.55 \text{ } m$$

$$X_{\text{NH}_4\text{Cl}} = \frac{\text{mol NH}_4\text{Cl}}{\text{total moles}} = \frac{2.324 \text{ mol}}{(50.60 + 2.324) \text{ mol}} = 0.043\underline{9}1 = 0.0439$$

12.49   In 1.00 mol of gas mixture, there are 0.35 mol of propane (pro.) and 0.65 mol of butane (but.). First calculate the masses of these components.

$$0.35 \text{ mol pro. } \times \frac{44.10 \text{ g pro.}}{1 \text{ mol pro.}} = 1\underline{5}.435 \text{ g pro.}$$

$$0.65 \text{ mol but. } \times \frac{58.12 \text{ g but.}}{1 \text{ mol but.}} = 3\underline{7}.778 \text{ g but.}$$

The mass of 1.00 mol of gas mixture is the sum of the masses of the two components:

15.435 g pro. + 37.778 g but. = 5\underline{3}.213 g mixture

Therefore, in 53.2 g of the mixture there are 15.4 g of propane and 37.8 g of butane. For a sample with a mass of 75 g:

$$75 \text{ g mixture } \times \frac{15.4 \text{ g pro.}}{53.2 \text{ g mixture}} = 2\underline{1}.7 = 22 \text{ g pro.}$$

$$75 \text{ g mixture } \times \frac{37.8 \text{ g pro.}}{53.2 \text{ g mixture}} = 5\underline{3}.2 = 53 \text{ g but.}$$

12.51   Calculate the moles of $KAl(SO_4)_2 \cdot 12H_2O$ using its molar mass of 474.4 g/mol, and use this to calculate the three concentrations.

a. The moles of $KAl(SO_4)_2 \cdot 12H_2O$ are calculated below using the abbreviation of "Hyd" for the formula of $KAl(SO_4)_2 \cdot 12H_2O$.

$$\text{mol Hyd. } = 0.9678 \text{ g Hyd. } \times \frac{1 \text{ mol Hyd}}{474.4 \text{ g Hyd}} = 0.00204\underline{0} = 0.002040 \text{ mol}$$

(continued)

Note that one mol of $KAl(SO_4)_2 \cdot 12H_2O$ contains one mole of $KAl(SO_4)_2$, so calculating the molarity of $KAl(SO_4)_2$ can be performed using the moles of $KAl(SO_4)_2 \cdot 12H_2O$.

$$\frac{mol}{L} = \frac{0.002040 \text{ mol Hyd}}{1.000 \text{ L soln.}} \times \frac{1 \text{ mol } KAl(SO_4)_2}{1 \text{ mol Hyd}} = 0.002040 \text{ M } KAl(SO_4)_2$$

b. The molarity of the SO_4^{2-} ion will be twice that of the $KAl(SO_4)_2$.

$$\frac{mol\ SO_4^{2-}}{L} = 0.002040 \text{ M } KAl(SO_4)_2 \times \frac{2 \text{ mol } SO_4^{2-}}{1 \text{ mol } KAl(SO_4)_2} = 0.004080 \text{ M}$$

c. Since the density of the solution is 1.00 g/mL, the mass of 1.000 L of solution is 1000 g, or 1.000 kg. Since molality is moles per 1.000 kg of solvent, the molality of $KAl(SO_4)_2$ equals 0.002040 moles divided by 1.000 kg or 0.002040 m (the same as the molarity).

12.53 In 1.00 kg of a saturated solution of urea, there are 0.44 kg of urea (a molecular solute) and 0.56 kg of water. First convert the mass of urea to moles.

$$0.44 \times 10^3 \text{ g urea} \times \frac{1 \text{ mol urea}}{60.06 \text{ g urea}} = 7.\underline{3}26 \text{ mol urea}$$

Then find the molality of the urea in the solution:

$$\text{Molality} = \frac{\text{mol urea}}{\text{kg } H_2O} \times \frac{7.326 \text{ mol urea}}{0.56 \text{ kg}} = 13.\underline{0}8 \text{ m}$$

$$\Delta T_f = K_f c_m = (1.858°C/m)(13.08 \text{ m}) = 24.\underline{3}°C$$

$$T_f = 0.0°C - 24.3°C = -2\underline{4}.3 = -24°C$$

12.55 $$M = \frac{\pi}{RT} = \frac{7.7 \text{ atm}}{(0.0821 \text{ L} \cdot \text{atm/K} \cdot \text{mol})(37 + 273)K} = 0.3\underline{0}2 = 0.30 \text{ mol/L}$$

12.57 Consider the equation $\Delta T_f = iK_f c_m$. For $CaCl_2$, $i = 3$; for glucose $i = 1$. The product of i and c_m for the solutions can be compared to determine which will have the larger ΔT_f and correspondingly lower freezing point. For $CaCl_2$, $ic_m = 3 \times 0.045 = 0.135$. For glucose $ic_m = 1 \times 0.15 = 0.15$. Therefore the solution of glucose will have the larger ΔT_f and the lower freezing point.

12.59   Assume there is 1.000 L of the solution, which will contain 18 mol H_2SO_4, molar mass 98.09 g/mol. The mass of the solution is

$$\text{Mass solution} = \frac{18 \text{ mol} \times 98.09 \text{ g/mol}}{0.98} = 1\underline{8}02 \text{ g}$$

Thus, the density of the solution is

$$d = \frac{1802 \text{ g}}{1000 \text{ mL}} = 1.\underline{8}02 = 1.8 \text{ g/mL}$$

The mass of water in the solution is

$$\text{mass } H_2O = 1802 \text{ g} \times 0.02 = \underline{3}6.0 \text{ g} = 0.0\underline{3}60 \text{ kg}$$

Thus, the molality of the solution is

$$m = \frac{18 \text{ mol } H_2SO_4}{0.0360 \text{ Kg}} = \underline{5}00 = 5 \times 10^2 \text{ } m$$

12.61   Use the freezing point depression equation to find the molality of the solution. The freezing point of pure cyclohexane is 6.55°C, and $K_f = 20.2°C/m$. Thus

$$m = \frac{\Delta T_f}{K_f} = \frac{(6.55 - 5.28)°C}{20.2°C/m} = 0.062\underline{8}7 \text{ } m$$

The moles of the compound are

$$\text{mol compound} = \frac{0.06287 \text{ mol}}{1 \text{ kg solvent}} \times 0.00538 \text{ kg} = 3.3\underline{8}2 \times 10^{-4} \text{ mol}$$

The molar mass of the compound is

$$\text{molar mass} = \frac{0.125 \text{ g}}{3.382 \times 10^{-4} \text{ mol}} = 36\underline{9}.6 \text{ g/mol}$$

The moles of the elements in 100 g of the compound are

$$\text{mol Mn} = 28.17 \text{ g Mn} \times \frac{1 \text{ mol}}{54.94 \text{ g Mn}} = 0.512\underline{7}4 \text{ mol}$$

(continued)

$$\text{mol C} = 30.80 \text{ g C} \times \frac{1 \text{ mol}}{12.01 \text{ g C}} = 2.5645 \text{ mol}$$

$$\text{mol O} = 41.03 \text{ g O} \times \frac{1 \text{ mol}}{16.00 \text{ g O}} = 2.5644 \text{ mol}$$

This gives mole ratios of one mol Mn to five mol C to five mol O. Therefore the empirical formula of the compound is MnC_5O_5. The weight of this formula unit is approximately 195 amu. Since the molar mass of the compound is 370. g/mol, the value of n is

$$n = \frac{370 \text{ g/mol}}{195 \text{ g/unit}} = 2.00, \text{ or } 2$$

Therefore the formula of the compound is $Mn_2C_{10}O_{10}$.

■ Solutions to Cumulative-Skills Problems

12.63

$Na^+(g) + Cl^-(g)$	$\rightarrow$	$NaCl(s)$	$\Delta H = -787$ kJ/mol
$NaCl(s)$	$\rightarrow$	$Na^+(aq) + Cl^-(aq)$	$\Delta H = +4$ kJ/mol
$Na^+(g) + Cl^-(g)$	$\rightarrow$	$Na^+(aq) + Cl^-(aq)$	$\Delta H = -783$ kJ/mol

The heat of hydration of Na^+ is

$Na^+(g) + Cl^-(g)$	$\rightarrow$	$Na^+(aq) + Cl^-(aq)$	$\Delta H = -783$ kJ/mol
$Cl^-(aq)$	$\rightarrow$	$Cl^-(g)$	$\Delta H = +338$ kJ/mol
$Na^+(g)$	$\rightarrow$	$Na^+(aq)$	$\Delta H = -445$ kJ/mol

12.65 $15.0 \text{ g MgSO}_4\text{•}7H_2O \times \dfrac{1 \text{ mol}}{246.5 \text{ g}} = 0.060854 \text{ mol MgSO}_4\text{•}7H_2O$

$0.060854 \text{ mol MgSO}_4\text{•}7H_2O \times \dfrac{7 \text{ mol } H_2O}{1 \text{ mol hydrate}} \times \dfrac{18.0 \text{ g } H_2O}{1 \text{ mol } H_2O} = 7.667 \text{ g } H_2O$

(continued)

$$\text{kg } H_2O = (100.0 \text{ g } H_2O + 7.667 \text{ g } H_2O) \times \frac{1 \text{ kg } H_2O}{1000 \text{ g } H_2O} = 0.10766 \text{ kg } H_2O$$

$$m = \frac{0.060854 \text{ mol } MgSO_4}{0.10766 \text{ kg } H_2O} = 0.5652 \text{ mol/kg} = 0.565 \text{ } m$$

12.67 $$15.0 \text{ g } CuSO_4 \cdot 5H_2O \times \frac{1 \text{ mol}}{249.7 \text{ g}} = 0.06007 \text{ mol } CuSO_4 \cdot 5H_2O, \text{ or } CuSO_4$$

$$100 \text{ g soln} \times \frac{1 \text{ mL soln}}{1.167 \text{ g soln}} \times \frac{1 \text{ L}}{1000 \text{ mL}} = 0.085689 \text{ L}$$

$$M = \frac{0.06007 \text{ mol } CuSO_4}{0.085689 \text{ L}} = 0.70102 = 0.701 \text{ mol/L}$$

12.69   Calculate the empirical formula first, using the masses of C, O, and H in 1.000 g:

$$1.434 \text{ g } CO_2 \times \frac{12.01 \text{ g C}}{44.01 \text{ g } CO_2} = 0.39132 \text{ g C}$$

$$0.783 \text{ g } H_2O \times \frac{2.016 \text{ g H}}{18.016 \text{ g } H_2O} = 0.08761 \text{ g H}$$

g O = 1.000 g - 0.39132 g - 0.08761 g = 0.5211 g O

Mol C = 0.39132 g C x 1 mol/12.01 g = 0.03258 mol C (lowest integer = 1)

Mol H = 0.08761 g H x 1 mol/1.008 g = 0.08691 mol H (lowest integer = 8/3)

Mol O = 0.5211 g O x 1 mol/16.00 g = 0.03257 mol O (lowest integer = 1)

Therefore the empirical formula is $C_3H_8O_3$. The formula weight from the freezing point is calculated by first finding the molality:

0.0894°C x (m/1.858°C) = 0.04811 = (0.04811 mol/1000 g H_2O)

(0.04811 mol/1000 g H_2O) x 25.0 g H_2O = 0.001203 mol (in 25.0 g H_2O)

Molar mass = M_m = 0.1107 g/0.001203 mol = 92.02 g/mol

Because this is also the formula weight, the molecular formula is also $C_3H_8O_3$.

13. RATES OF REACTION

■ Answers to Review Questions

13.1 The four variables that can affect rate are (1) the concentrations of the reactants, although in some cases a particular reactant's concentration does not affect the rate; (2) the presence and concentration of a catalyst; (3) the temperature of the reaction; and (4) the surface area of any solid reactant or solid catalyst.

13.2 The rate of reaction of HBr can be defined as the decrease in HBr concentration (or the increase in Br_2 product formed) over the time interval, Δt:

$$\text{Rate} = -1/4 \, \frac{\Delta[HBr]}{\Delta t} = 1/2 \, \frac{\Delta[Br_2]}{\Delta t} \quad \text{or} \quad -\frac{\Delta[HBr]}{\Delta t} = 2 \, \frac{\Delta[Br_2]}{\Delta t}$$

13.3 Two physical properties used to determine the rate are color, or absorption of electromagnetic radiation, and pressure. If a reactant or product is colored or absorbs a different type of electromagnetic radiation than the other species, then measurement of the change in color (change in absorption of electromagnetic radiation) may be used to determine the rate. If a gas reaction involves a change in the number of gaseous molecules, measurement of the pressure change may be used to determine the rate.

13.4 The rate law for this reaction of iodide ion, arsenic acid, and hydrogen ion is

$$\text{Rate} = k[I^-][H_3AsO_4][H^+]$$

The overall order is $1 + 1 + 1 = 3$ (third order).

13.5 Use m to symbolize the reaction order as is done in the text. Then from the table for m and the change in rate in the text, m is two when the rate is quadrupled (increased fourfold). Using the equation in the text gives the same result:

$$2^m = \text{new rate/old rate} = 4/1; \text{ thus, } m = 2$$

13.6   Use m to symbolize the reaction order as is done in the text. The table for m and the change in rate in the text cannot be used in this case. When $m = 0.5$, the new rate should be found using the equation in the text:

$$2^{0.50} = \sqrt{2} = 1.41 = \text{new rate/old rate}$$

Thus, the new rate is 1.41 times the old rate.

13.7   Use the half-life concept to answer the question without an equation. If the half-life for the reaction of A(g) is 25 s, then the time for A(g) to decrease to 1/4 the initial value is two half-lives or 2 x 25 = 50 s. The time for A(g) to decrease to 1/8 the initial value is three half-lives, or 3 x 25 = 75 s.

13.8   According to transition-state theory, the two factors that determine whether a collision results in reaction or not are (1) the molecules must collide with the proper orientation to form the activated complex, and (2) the activated complex formed must have a kinetic energy greater than the activation energy.

13.9   The potential-energy diagram for the exothermic reaction of A and B to give activated complex AB‡ and products C and D is given below.

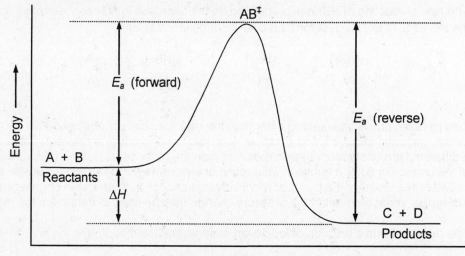

13.10 The Arrhenius equation expressed with the base e is

$$k = A e^{-E_a/RT}$$

The A term is the frequency factor and is equal to the product of p and Z from collision theory. The term p is the fraction of collisions with properly oriented reactant molecules, and Z is the frequency of collisions. Thus A is the number of collisions with the molecules properly oriented. The E_a term is the activation energy, the minimum energy of collision required for two molecules to react. The R term is the gas constant, and T is the absolute temperature.

13.11 In the reaction of $NO_2(g)$ with $CO(g)$, an example of an intermediate is the temporary formation of NO_3 from the reaction of two NO_2 molecules in the first step:

$$NO_2 + NO_2 \rightarrow NO_3 + NO$$

$$NO_3 + CO \rightarrow NO_2 + CO_2$$

13.12 It is generally impossible to predict the rate law from the equation alone because most reactions consist of several elementary steps whose combined result is summarized in the rate law. If these elementary steps are unknown, the rate law cannot be predicted.

13.13 For the rate of decomposition of N_2O_4,

$$Rate = k_1[N_2O_4]$$

For the rate of formation of N_2O_4,

$$Rate = k_{-1}[NO_2]^2$$

At equilibrium the rates are equal, so

$$k_1[N_2O_4] = k_{-1}[NO_2]^2$$

$$[N_2O_4] = (k_{-1}/k_1) [NO_2]^2$$

13.14 A catalyst operates by providing a pathway (mechanism) that occurs faster than the uncatalyzed pathway (mechanism) of the reaction. The catalyst is not consumed because, after reacting in an early step, it is regenerated in a later step.

13.15 In physical adsorption, molecules adhere to a surface through *weak* intermediate forces, whereas in chemisorption the molecules adhere to the surface by *stronger* chemical bonding.

13.16   In the first step of catalytic hydrogenation of ethylene, the ethylene and hydrogen molecules diffuse to the catalyst surface and undergo chemisorption. Then the pi electrons of ethylene form temporary bonds to the metal catalyst, and the hydrogen molecule breaks into two hydrogen atoms. The hydrogen atoms next migrate to an ethylene held in position on the metal catalyst surface, forming ethane. Finally, because it cannot bond to the catalyst, the ethane diffuses away from the surface.

■ Solutions to Practice Problems

Note on significant figures: If the final answer to a solution needs to be rounded off, it is given first with one nonsignificant figure, and the last significant figure is underlined. The final answer is then rounded to the correct number of significant figures. In multiple-step problems, intermediate answers are given with at least one nonsignificant figure; however, only the final answer has been rounded off.

13.23   For the reaction $2NO_2 \rightarrow 2NO + O_2$, the rate of decomposition of NO_2 and the rate of formation of O_2 are, respectively,

$$\text{Rate} = -\Delta[NO_2]/\Delta t$$

$$\text{Rate} = \Delta[O_2]/\Delta t$$

To relate the two rates, divide each rate by the coefficient of the corresponding substance in the chemical equation, and equate them.

$$-1/2 \frac{\Delta[NO_2]}{\Delta t} = \frac{\Delta[O_2]}{\Delta t}$$

13.25 $$\text{Rate} = -\frac{\Delta[NH_4NO_2]}{\Delta t} = -\frac{[0.0432\ M - 0.500\ M]}{[3.00\ hr - 0.00\ hr]} = 2.2\underline{7} \times 10^{-2} = 2.3 \times 10^{-2}\ M/hr$$

13.27 $$\text{Rate} = -\frac{\Delta[Azo.]}{\Delta t} = -\frac{[0.0101\ M - 0.0150\ M]}{7.00\ min - 0.00\ min} \times \frac{1\ min}{60\ sec} = 1.\underline{16} \times 10^{-5}$$

$$= 1.2 \times 10^{-5}\ M/s$$

13.29   If the rate law is rate = $k[MnO_4^-][H_2C_2O_4]$, the order with respect to MnO_4^- is one (first-order), the order with respect to $H_2C_2O_4$ is one (first-order), and the order with respect to H^+ is zero. The overall order is two, second-order.

13.31 The reaction rate doubles when the concentration of CH_3NNCH_3 is doubled, so the reaction is first-order in azomethane. The rate equation should have the form

Rate $= k[CH_3NNCH_3]$

Substituting values for the rate and concentration yields a value for k:

$$k = \frac{\text{rate}}{[\text{Azo.}]} = \frac{2.8 \times 10^{-6} \ M/s}{1.13 \times 10^{-2} \ M} = 2.4\underline{7} \times 10^{-4} = 2.5 \times 10^{-4}/s$$

13.33 Doubling [NO] quadruples the rate, so the reaction is second-order in NO. Doubling $[H_2]$ doubles the rate, so the reaction is first-order in H_2. The rate law should have the form

Rate $= k[NO]^2[H_2]$

Substituting values for the rate and concentrations yields a value for k:

$$k = \frac{\text{rate}}{[NO]^2[H_2]} = \frac{2.6 \times 10^{-5} \ M/s}{[6.4 \times 10^{-3} \ M]^2[2.2 \times 10^{-3} \ M]} = 2.8\underline{8} \times 10^2 = 2.9 \times 10^2 \ M^2 s$$

13.35 First, find the rate constant, k, by substituting experimental values into the first-order rate equation. Let $[\text{Et. Cl.}]_o = 0.00100 \ M$, $[\text{Et. Cl.}]_t = 0.00067 \ M$, and $t = 155$ s. Solving for k yields

$$k = - \frac{\ln \dfrac{[0.00067 \ M]_t}{[0.00100 \ M]_o}}{155 \ s} = 2.5\underline{8}3 \times 10^{-3}/s$$

Now let $[\text{Et. Cl.}]_t =$ the concentration after 256 s, $[\text{Et. Cl.}]_o$ again $= 0.00100 \ M$, and use the value of k of $1.5\underline{6}4 \times 10^{-3}/s$ to calculate $[\text{Et. Cl.}]_t$.

$$\ln \frac{[\text{Et. Cl.}]_t}{[0.00100 \ M]} = - (2.584 \times 10^{-3}/s)(256 \ s) = -0.6\underline{6}14$$

Converting both sides to antilogs gives

$$\frac{[\text{Et. Cl.}]_t}{[0.00100 \ M]} = 0.5\underline{1}61$$

$[\text{Et. Cl.}]_t = 0.5161 \times [0.00100 \ M] = 5.\underline{1}6 \times 10^{-4} = 5.2 \times 10^{-4} \ M$

13.37 For a first-order reaction, the rate law can be rearranged and solved for time to get the following relation.

$$t = -\frac{1}{k} \ln \frac{[A]_t}{[A]_o}$$

The half-life is

$$t_{1/2} = \frac{0.693}{k} = \frac{0.693}{6.3 \times 10^{-4}/s} = 1.\underline{1}0 \times 10^3 = 1.1 \times 10^3 \, s$$

The time it takes for the concentration to decrease to 75.0% of its original value is

$$t = -\frac{1}{6.3 \times 10^{-4}/s} \ln[.75] = 4\underline{5}6 = 4.6 \times 10^2 \, s \, (7.6 \, min)$$

The time it takes for the concentration to decrease to 30.0% of its original value is

$$t = -\frac{1}{6.3 \times 10^{-4}/s} \ln[.30] = 1\underline{9}11 = 1.9 \times 10^3 \, s \, (32 \, min)$$

13.39 For a first-order reaction, divide 0.693 by the rate constant to find the half-life:

$$t_{1/2} = 0.693/(2.0 \times 10^{-6}/s) = 3.\underline{4}65 \times 10^5 \, s \, (9\underline{6}.25 \text{ or } 96 \, hr)$$

$$t_{25.0\% \text{ left}} = t_{1/4 \text{ left}} = 2 \times t_{1/2} = 2 \times 96.25 \, hr = 1\underline{9}2.5 = 1.9 \times 10^2 \, hr$$

$$t_{12.5\% \text{ left}} = t_{1/8 \text{ left}} = 3 \times t_{1/2} = 3 \times 96.25 \, hr = 2\underline{8}8.75 = 2.9 \times 10^2 \, hr$$

$$t_{6.25\% \text{ left}} = t_{1/16 \text{ left}} = 4 \times t_{1/2} = 4 \times 96.25 \, hr = 3\underline{8}5.0 = 3.9 \times 10^2 \, hr$$

$$t_{3.125\% \text{ left}} = t_{1/32 \text{ left}} = 5 \times t_{1/2} = 5 \times 96.25 \, hr = 4\underline{8}1.25 = 4.8 \times 10^2 \, hr$$

13.41 For the first-order plot, follow Figure 13.6, and plot $\ln[ClO_2]$ versus the time in seconds. The data used for plotting are

t. min	$\ln[ClO_2]$
0.00	-7.647
1.00	-7.749
2.00	-7.846
3.00	-7.949

(continued)

The plot yields an approximate straight line, demonstrating that the reaction is first-order in $[ClO_2]$. The slope of the line may be calculated from the difference between the last point and the first point:

$$Slope = \frac{[(-7.949) - (-7.647)]}{[3.00 - 0.00]s} = -0.10\underline{0}6/s$$

Just as the slope, m, was obtained for the plot in Figure 13.6, you can also equate m to $-k$ and calculate k as follows:

$$k = -slope = 0.10\underline{0}6 = 0.101/s$$

13.43 The rate law for a zero-order reaction is

$$[A] = -kt + [A]_o.$$

Using a rate constant of 8.1×10^{-2} mol/(L•s) and an initial concentration of 0.10 M, the time it would take for the concentration to change to 1.0×10^{-2} M is

$$1.0 \times 10^{-2}\ M = -8.1 \times 10^{-2}\ mol/(L•s) \times t + 0.10\ M$$

$$t = \frac{0.10\ M - 1.0 \times 10^{-2}\ M}{8.1 \times 10^{-2}\ M/s} = 1.\underline{1}1 = 1.1\ s$$

13.45 The potential-energy diagram is below. Because the activation energy for the forward reaction is +10 kJ, and $\Delta H° = -200$ kJ, the activation energy for the reverse reaction is +210 kJ.

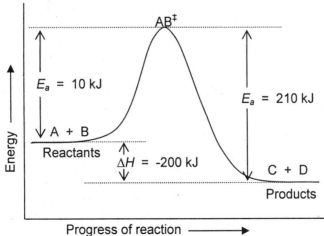

13.47 Solve the two-temperature Arrhenius equation for E_a by substituting $T_1 = 308$ K (from 35°C), $k_1 = 1.4 \times 10^{-4}$/s, $T_2 = 318$ K (from 45°C), and $k_2 = 5.0 \times 10^{-4}$/s:

$$\ln \frac{5.0 \times 10^{-4}}{1.4 \times 10^{-4}} = \frac{E_a}{8.31 \ \text{J/(mol} \cdot K)} \left(\frac{1}{308 \ \text{K}} - \frac{1}{318 \ \text{K}} \right)$$

Rearranging E_a to the left side and calculating [1/308 - 1/318] gives

$$E_a = \frac{8.31 \ \text{J/K} \times \ln(3.5714)}{1.0209 \times 10^{-4}/\text{K}} = 1.0\underline{3}7 \times 10^5 = 1.0 \times 10^5 \ \text{J/mol}$$

To find the rate at 75°C (348 K), use the first equation. Let k_2 in the numerator be the unknown and solve:

$$\ln \frac{k_2}{1.4 \times 10^{-4}} = \frac{1.037 \times 10^5 \ \text{J/mol}}{8.31 \ \text{J/(mol} \cdot K)} \left(\frac{1}{308 \ \text{K}} - \frac{1}{348 \ \text{K}} \right) = 4.\underline{6}57$$

$$\frac{k_2}{1.4 \times 10^{-4}} = e^{4.657} = 1\underline{0}5.3$$

Solving for k_2 gives

$$k_2 = 11.80 \times (1.4 \times 10^{-4}/\text{s}) = 0.01\underline{4}7 = 0.015/\text{s}$$

13.49 For plotting ln k versus $1/T$, the data below are used:

k	ln k	$1/T$ (1/K)
0.527	-0.64055	1.686×10^{-3}
0.776	-0.25360	1.658×10^{-3}
1.121	0.11422	1.631×10^{-3}
1.607	0.47436	1.605×10^{-3}

The plot yields an approximate straight line. The slope of the line is calculated from the difference between the last and the first points:

$$\text{Slope} = \frac{(0.47436) - (-0.64055)}{[1.605 \times 10^{-3} - 1.686 \times 10^{-3}]/\text{K}} = -137\underline{6}4 \ \text{K}$$

Because the slope = $-E_a/R$, you can solve for E_a using $R = 8.31$ J/K:

$$\frac{-E_a}{8.31 \ \text{J/(K} \cdot \text{mol)}} = -13764 \ \text{K}$$

$$E_a = 5976 \ \text{K} \times 8.31 \ \text{J/K} = 1.\underline{1}43 \times 10^5 \ \text{J/mol, or } 1.1 \times 10^2 \ \text{kJ/mol}$$

13.51   The $NOCl_2$ is a reaction intermediate that is produced in the first reaction and consumed in the second. The overall reaction is the sum of the two elementary reactions:

$$NO \quad + \; Cl_2 \; \rightarrow \quad NOCl_2$$

$$NOCl_2 \; + \; NO \; \rightarrow \quad 2NOCl$$

$$\overline{2NO \quad + \; Cl_2 \; \rightarrow \quad 2NOCl}$$

13.53   Step 1 of the isomerization of cyclopropane, C_3H_6, is slow, so the rate law for the overall reaction will be the rate law for this step, with $k_1 = k$, the overall rate constant:

Rate $= k[C_3H_6]^2$

13.55   Step 2 of this reaction is slow, so the rate law for the overall reaction would appear to be the rate law for this step:

Rate $= k_2[I]^2[H_2]$

However, the rate law includes an intermediate, the I atom, and cannot be used unless the intermediate is eliminated. This can only be done using an equation for step 1. At equilibrium, you can write the following equality for step 1:

$$k_1[I_2] \; = \; k_{-1} \, [I]^2$$

Rearranging and then substituting for the $[I]^2$ term yields

$$[I]^2 \; = \; [I_2] \frac{k_1}{k_{-1}}$$

Rate $= k_2(k_1/k_{-1})[I_2][H_2] \; = \; k[I_2][H_2]$ ($k =$ the overall rate constant)

■ Solutions to General Problems

13.57 All rates of reaction are calculated by dividing the decrease in concentration by the difference in times; hence, only the setup for the first rate (after 10 minutes) is given below. This setup is

$$\text{Rate (10 min)} = -\frac{(1.29 - 1.50) \times 10^{-2} \; M}{(10 - 0) \text{ min}} \times \frac{1 \text{ min}}{60 \text{ s}}$$

$$= 3.5 \times 10^{-6} \; M/s$$

A summary of the times and rates is given in the table.

Time, min	Rate
10	$3.\underline{5}0 \times 10^{-6} = 3.5 \times 10^{-6}$ M/s
20	$3.\underline{1}7 \times 10^{-6} = 3.2 \times 10^{-6}$ M/s
30	$2.\underline{5}0 \times 10^{-6} = 2.5 \times 10^{-6}$ M/s

13.59 The second-order integrated rate law is

$$\frac{1}{[A]_t} = kt + \frac{1}{[A]_o}$$

The starting concentration of NO_2 (A) is 0.050 M. Using a rate constant of 0.775 L/(mol•s), after 2.5×10^2 s the concentration of NO_2 will be

$$\frac{1}{[A]_t} = 0.775 \text{ L/(mol•s)}(2.5 \times 10^2 \text{ s}) + \frac{1}{0.050 \text{ mol/L}}$$

$$\frac{1}{[A]_t} = 19\underline{3} \text{ L/mol} + 2\underline{0}.0 \text{ L/mol} = 21\underline{3}. \text{ L/mol}$$

$$[A]_t = \frac{1}{213 \text{ L/mol}} = 4.\underline{6}7 \times 10^{-3} = 4.7 \times 10^{-3} \; M$$

13.61 First find the rate constant from the first-order rate equation, substituting the initial concentration of $[comp.]_o = 0:0350$ M and the $[comp.]_t = 0.0250$ M.

$$\ln \frac{0.0250\ M}{0.0350\ M} = -k\,(75\ s)$$

Rearranging and solving for k gives

$$k = -\frac{\left(\ln \dfrac{0.0250\ M}{0.0350\ M}\right)}{75\ s} = 4.\underline{4}8 \times 10^{-3}/s$$

Now arrange the first-order rate equation to solve for $[comp.]_t$; substitute the above value of k, again using $[comp.]_o = 0.0350$ M.

$$\ln \frac{[comp.]_t}{0.0350\ M} = -(4.48 \times 10^{-3}/s)(95\ s) = -0.4\underline{2}61$$

Taking the antilog of both sides gives

$$\frac{[comp.]_t}{0.0350\ M} = e^{-0.4261} = 0.6\underline{5}29$$

$$[comp]_t = 0.6529 \times [0.0350\ M]_o = 0.02\underline{2}8 = 0.023\ M$$

13.63 The $\ln[CH_3NNCH_3]$ and time data for the plot are tabulated below.

$t.$ min	$\ln[CH_3NNCH_3]$
0	-4.1997
10	-4.3505
20	-4.5098
30	-4.6564

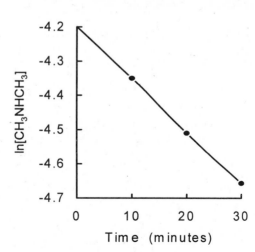

From the graph the slope, m, is calculated:

$$m = \frac{(-4.6564) - (-4.1997)}{(30 - 0)\ \text{min}}$$

$$= -0.01\underline{5}22\ /\text{min}$$

Because the slope also $= -k$, this gives

$$k = -(-0.01522/\text{min})(1\ \text{min}/60\ s) = 2.\underline{5}3 \times 10^{-4} = 2.5 \times 10^{-4}/s$$

13.65   The rate law for a zero-order reaction is

$$[A] = -kt + [A]_o.$$

Using a rate constant of 3.7×10^{-6} mol/(L•s) and an initial concentration of 5.0×10^{-4} M, the time it would take for the concentration to drop to 5.0×10^{-5} M is

$$5.0 \times 10^{-5} \, M = -8.1 \times 10^{-2} \text{ mol/(L•s)} \times t + 5.0 \times 10^{-4} \, M$$

$$t = \frac{(5.0 \times 10^{-4} \, M) - (5.0 \times 10^{-5} \, M)}{3.7 \times 10^{-6} \, M/s} = 121.6 = 1.2 \times 10^2 \text{ s}$$

13.67   The slow step determines the observed rate, so the overall rate constant, k, should be equal to the rate constant for the first step, and the rate law should be

Rate = $k[NO_2Br]$

13.69   The slow step determines the observed rate; assuming k_2 is the rate constant for the second step, the rate law would appear to be

Rate = $k_2[NH_3][HOCN]$

However, this rate law includes two intermediate substances that are neither reactants nor products. This rate law cannot be used unless both are eliminated. This can only be done using an equation from step 1. At equilibrium in step 1, you can write the following equality, assuming k_1 and k_{-1} are the rate constants for the forward and back reactions, respectively:

$k_1[NH_4^+][OCN^-] = k_{-1}[NH_3][HOCN]$

Rearranging and then substituting for the $[NH_3][HOCN]$ product gives

$[NH_3][HOCN] = (k_1/k_{-1})[NH_4^+][OCN^-]$

Rate = $k_2(k_1/k_{-1})[NH_4^+][OCN^-] = k[NH_4^+][OCN^-]$ (k = overall rate constant)

13.71 a.   The reaction is first-order in O_2 because the rate doubled with a doubling of the oxygen concentration. The reaction is second-order in NO because the rate increased by a factor of eight when both the NO and O_2 concentrations were doubled.

Rate = $k[NO]^2[O_2]$

(continued)

b. The initial rate of the reaction for Experiment 4 can be determined by first calculating the value of the rate constant using Experiment 1 for the data.

0.80×10^{-2} M/s $= k[4.5 \times 10^{-2}$ M$]^2[2.2 \times 10^{-2}$ M$]$

Solving for the rate constant gives

$k = 1.\underline{7}96 \times 10^2$ $M^{-2}s^{-1}$

Now, use the data in Experiment 4 and the rate constant to determine the initial rate of the reaction.

Rate $= 1.796 \times 10^2$ $M^{-2}s^{-1}$ $[3.8 \times 10^{-1}$ M$]^2[4.6 \times 10^{-3}$ M$]$

Rate $= 0.1\underline{1}9$ $= 0.12$ mol/L•s

13.73   The diagram:

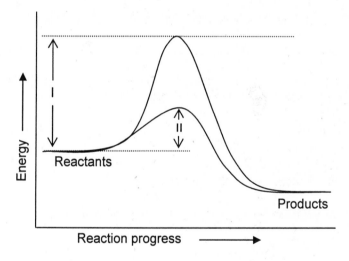

In the diagram, I represents the activation energy for the uncatalyzed reaction, and II represents the activation energy for the catalyzed reaction. A catalyst provides another pathway for a chemical reaction and, with a lower activation energy, more molecules have enough energy to react, so the reaction will be faster.

13.75 The rate of a chemical reaction is the change in the concentration of a reactant or product with time. For a reactant,

$-\Delta c/\Delta t$ or $-d[c]/dt$.

The rate changes because the concentration of the reactant has changed.

Rate = $k[A]^m$

The rate will equal k when the reactants all have 1.00 M concentrations.

Rate = $k[A]^m[B]^n$

■ Solutions to Cumulative-Skills Problems

13.77 Using the first-order rate law, the initial rate of decomposition is given by

Rate = $k[H_2O_2]$ = $(7.40 \times 10^{-4}/s) \times (1.50\ M\ H_2O_2)$ = $1.110 \times 10^{-3}\ M\ H_2O_2/s$

The heat liberated per second per mol of H_2O_2 can be found by first calculating the standard enthalpy of the decomposition of one mol of H_2O_2:

$H_2O_2(aq)$	→	$H_2O(l)$	+	1/2 $O_2(g)$
$\Delta H°_f$ = -191.2 kJ/mol		-285.84 kJ/mol		0 kJ/mol

For the reaction, the standard enthalpy change is

$\Delta H°$ = -285.84 kJ/mol - (-191.2 kJ/mol) = -94.64 kJ/mol H_2O_2

The heat liberated per second is:

$$\frac{94.64\ kJ}{mol\ H_2O_2} \times \frac{1.110 \times 10^{-3}\ mol\ H_2O_2}{L \cdot s} \times 2.00\ L = 0.21010 = 0.210\ kJ/s$$

13.79 Use the ideal gas law ($P/RT = n/V$) to calculate the mol/L of each gas:

$$[O_2] = \frac{(345/760)\ atm}{0.082057\ L \cdot atm/(K \cdot mol) \times 612\ K} = 0.009039\ mol\ O_2/L$$

$$[NO] = \frac{(155/760)\ atm}{0.082057\ L \cdot atm/(K \cdot mol) \times 612\ K} = 0.004061\ mol\ NO/L$$

The rate of decrease of NO is

$$\frac{1.16 \times 10^{-5}\ L^2}{mol^2 \cdot s} \times \left(\frac{4.061 \times 10^{-3}\ mol}{1\ L}\right)^2 \times \frac{9.039 \times 10^{-3}\ mol}{1\ L}$$

$$= 1.729 \times 10^{-12}\ mol/(L \cdot s)$$

The rate of decrease in atm/s is found by multiplying by RT:

$$\frac{1.729 \times 10^{-12}\ mol}{L \cdot s} \times \frac{0.082057\ L \cdot atm}{K \cdot mol} \times 612\ K = 8.682 \times 10^{-11}\ atm/s$$

The rate of decrease in mmHg/s is

$$8.682 \times 10^{-11}\ atm/s \times (760\ mmHg/atm) = 6.598 \times 10^{-8} = 6.60 \times 10^{-8}\ mmHg/s$$

14. CHEMICAL EQUILIBRIUM

■ Answers to Review Questions

14.1 A reasonable graph showing the decrease in concentration of $N_2O_4(g)$ and the increase in concentration of $NO_2(g)$ is shown below:

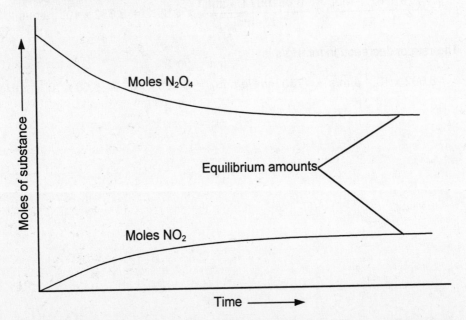

At first, the concentration of N_2O_4 is large, and the rate of the forward reaction is large, but then as the concentration of N_2O_4 decreases, the rate of the forward reaction decreases. In contrast, the concentration of NO_2 builds up from zero to a low concentration. Thus, the initial rate of the reverse reaction is zero, but it steadily increases as the concentration of NO_2 increases. Eventually, the two rates become equal when the reaction reaches equilibrium. This is a dynamic equilibrium because both the forward and reverse reactions are occurring at all times even though there is no net change in concentration at equilibrium.

14.2   The 1.0 mol of $H_2(g)$ and 1.0 mol of $I_2(g)$ in the first mixture reach equilibrium when the amounts of reactants decrease to 0.50 mol each and when the amount of product increases to 1.0 mol. The total number of moles of the reactants at the start is 2.0 mol, which is the same number of moles as in the second mixture, the 2.0 mol of HI that is to be allowed to come to equilibrium. The second mixture should produce the same number of moles of H_2, I_2, and HI at equilibrium because, if the total number of moles is constant, it should not matter from which direction an equilibrium is approached.

14.3   The addition of reactions 1 and 2 yields reaction 3:

(Reaction 1)	HCN	+ OH⁻	⇌	CN⁻	+	H_2O
(Reaction 2)		H_2O	⇌	H^+	+	OH⁻
(Reaction 3)		HCN	⇌	H^+	+	CN⁻

The rule states that, if a given equation can be obtained from the sum of other equations, the equilibrium constant for the given equation equals the product of the other equilibrium constants. Thus, K for reaction 3 is

$$K = K_1 \times K_2 = (4.9 \times 10^4) \times (1.0 \times 10^{-14}) = 4.9 \times 10^{-10}$$

14.4   a. Homogeneous equilibrium. All substances are gases and thus exist in one phase, a mixture of gases.

b. Heterogeneous equilibrium. The two copper compounds are solids, but the other substances are gases. This fulfills the definition of a heterogeneous equilibrium.

c. Homogeneous equilibrium. All substances are gases and thus exist in one phase, a mixture of gases.

d. Heterogeneous equilibrium. The two copper-containing substances are solids, but the other substances are gases. This fulfills the definition of a heterogeneous equilibrium.

14.5   Pure liquids and solids can be ignored in an equilibrium-constant expression because their concentrations do not change. (If a solid is present, it has not dissolved in any gas or solution present.) In effect, concentrations of liquids and solids are incorporated into the value of K_c, as discussed in the chapter.

14.6   A qualitative interpretation of the equilibrium constant involves using the magnitude of the equilibrium constant to predict the relative amounts of reactants and products at equilibrium. If K_c is around one, the equilibrium mixture contains appreciable amounts (same order of magnitude) of reactants and products. If K_c is large, the equilibrium mixture is mostly products. If K_c is small, the equilibrium mixture is mostly reactants. The type of reaction governs what "large" and "small" values are; but for some types of reactions, "large" might be no less than 10^2 to 10^4, whereas "small" might be no more than 10^{-4} to 10^{-2}.

14.7   In some cases, a catalyst can affect the product in a reaction because it affects only the rate of one reaction out of several reactions that are possible. If two reactions are possible, and the uncatalyzed rate of one is much slower but is the only reaction that is catalyzed, then the products with and without a catalyst will be different. The Ostwald process is a good example. In the absence of a catalyst, NH_3 burns in O_2 to form only N_2 and H_2O, even though it is possible for NH_3 to react to form NO and H_2O. Ostwald found that adding a platinum catalyst favors the formation of the NO and H_2O almost to the exclusion of the N_2 and H_2O.

14.8   Four ways in which the yield of ammonia can be improved are (1) removing the gaseous NH_3 from the equilibrium by liquefying it, (2) increasing the nitrogen or hydrogen concentration, (3) increasing the total pressure on the mixture (the moles of gas decrease), and (4) lowering the temperature ($\Delta H°$ is negative, so heat is evolved). Each causes a shift to the right in accordance with Le Châtelier's principle.

■ Solutions to Practice Problems

Note on significant figures: If the final answer to a solution needs to be rounded off, it is given first with one nonsignificant figure, and the last significant figure is underlined. The final answer is then rounded to the correct number of significant figures. In multiple-step problems, intermediate answers are given with at least one nonsignificant figure; however, only the final answer has been rounded off.

14.15   Because the amount of NO_2 at the start is zero, the 0.90 mol of NO_2 at equilibrium must also equal the change, x, in the moles of N_2O_3. Use the table approach, and insert the starting, change, and equilibrium number of moles:

Amt. (mol)	$N_2O_3(g)$	$\rightleftharpoons$	$NO_2(g)$	+	$NO(g)$
Starting	3.00		0		0
Change	-x		+x		+x
Equilibrium	3.00 - x		x = 0.90		x

Since the equilibrium amount of NO_2 is given in the problem, this tells you $x = 0.90$. The equilibrium amounts for the other substances can now be determined.

Equilibrium amount NO $= x = 0.90$ mol

Equilibrium amount $N_2O_3 = 3.00 - x = 3.00 - 0.90 = 2.10$ mol

Therefore, the amounts of the substances in the equilibrium mixture are 2.10 mol N_2O_3, 0.90 mol NO_2, and 0.90 mol NO.

14.17   Use the table approach, and give the starting, change, and equilibrium number of moles of each.

Amt. (mol)	2NO(g)	+	Br$_2$(g)	⇌	2NOBr(g)
Starting	0.0524		0.0262		0
Change	-2x		-x		+2x
Equilibrium	0.0524 - 2x		0.0262 - x		2x = 0.0311

Since the equilibrium amount of NOBr is given in the problem, this tells you x = 0.01555. The equilibrium amounts for the other substances can now be determined.

Equilibrium amount NO = 0.0524 - 2 x (0.01555) = 0.02130 = 0.0213 mol

Equilibrium amount Br$_2$ = 0.0262 - 0.01555 = 0.01065 = 0.0107 mol

Therefore, the amounts of the substances in the equilibrium mixture are 0.0213 mol NO, 0.0107 mol Br$_2$, and 0.0311 mol NOBr.

14.19 a. $K_c = \dfrac{[NO_2][NO]}{[N_2O_3]}$

b. $K_c = \dfrac{[H_2]^2[S_2]}{[H_2S]^2}$

c. $K_c = \dfrac{[NO_2]^2}{[NO]^2[O_2]}$

14.21   When the reaction is halved, the equilibrium-constant expression is

$$K_c = \frac{[NO_2]^2[O_2]^{1/2}}{[N_2O_5]}$$

When the reaction is then reversed, the equilibrium-constant expression becomes

$$K_c = \frac{[N_2O_5]}{[NO_2]^2[O_2]^{1/2}}$$

14.23 First calculate the molar concentrations of each of the compounds in the equilibrium:

$$[PCl_5] = 0.0126 \text{ mol } PCl_5 \div 4.00 \text{ L} = 0.003150 \text{ } M$$

$$[PCl_3] = 0.0148 \text{ mol } PCl_3 \div 4.00 \text{ L} = 0.003700 \text{ } M$$

$$[Cl_2] = 0.0870 \text{ mol } Cl_2 \div 4.00 \text{ L} = 0.02175 \text{ } M$$

Now substitute these into the expression for K_c:

$$K_c = \frac{0.003150}{(0.003700)(0.02175)} = 39.14 = 39.1$$

14.25 Substitute the following concentrations into the K_c expression:

$$[CH_3OH] = 0.0626 \text{ mol} \div 3.00 \text{ L} = 0.02086 \text{ } M$$

$$[CO] = 0.238 \text{ mol} \div 3.00 \text{ L} = 0.07933 \text{ } M$$

$$[H_2] = 0.474 \text{ mol} \div 3.00 \text{ L} = 0.1580 \text{ } M$$

$$K_c = \frac{(0.02086)}{(0.07933)(0.1580)^2} = 10.53 = 10.5$$

14.27 For each mole of NOBr that reacts, (1.000 - 0.094 = 0.906) mol remains. Starting with 2.00 mol NOBr, 2 x 0.906 mol NOBr, or 1.812 mol NOBr, remains. Because the volume is 1.00 L, the concentration of NOBr at equilibrium is 1.812 M. Assemble a table of starting, change, and equilibrium concentrations:

Conc. (M)	2NOBr(g) ⇌	2NO(g) +	Br$_2$(g)
Starting	2.00	0	0
Change	-2x	+2x	+x
Equilibrium	2.00 - 2x (= 1.812)	2x	x

Because 2.00 - 2x = 1.812, x = 0.094 M. Therefore, the equilibrium concentrations are [NOBr] = 1.812 M, [NO] = 0.188 M, and [Br$_2$] = 0.094 M.

$$K_c = \frac{[NO]^2[Br_2]}{[NOBr]^2} = \frac{(0.188 \text{ } M)^2(0.094 \text{ } M)}{(1.812 \text{ } M)^2} = 1.01 \times 10^{-3} = 1.0 \times 10^{-3}$$

14.29 a. $K_p = \dfrac{P_{HBr}^2}{P_{H_2} P_{Br_2}}$

b. $K_p = \dfrac{P_{CH_4} P_{H_2S}^2}{P_{CS_2} P_{H_2}^4}$

c. $K_p = \dfrac{P_{H_2O}^2 P_{Cl_2}^2}{P_{HCl}^4 P_{O_2}}$

14.31 For each one mol of gaseous product, there is 1.5 mol of gaseous reactants; thus $\Delta n = 1 - 1.5 = -0.5$. Using this, calculate K_c from K_p:

$$K_c = \frac{K_p}{(RT)^{\Delta n}} = \frac{6.55}{(0.0821 \times 900)^{-0.5}} = 6.55 \times (0.0821 \times 900)^{0.5}$$

$$= 56.\underline{3}03 = 56.3$$

14.33 a. $K_c = \dfrac{[CO]^2}{[CO_2]}$

b. $K_c = \dfrac{[CO_2]}{[CO]}$

c. $K_c = \dfrac{[CO_2]}{[SO_2][O_2]^{1/2}}$

14.35 a. Not complete; K_c is very small (10^{-31}), indicating very little reaction.

b. Nearly complete; K_c is very large (10^{21}), indicating nearly complete reaction.

14.37   Calculate Q, the reaction quotient, and compare it to the equilibrium constant ($K_c = 3.07 \times 10^{-4}$). If Q is larger, the reaction will go to the left; if smaller, the reaction will go to the right; if equal, the reaction is at equilibrium. In all cases, Q is found by combining these terms:

$$Q = \frac{[NO]^2\,[Br_2]}{[NOBr]^2}$$

a. $Q = \dfrac{(0.0162)^2(0.0123)}{(0.0720)^2} = 6.2\underline{2}6 \times 10^{-4} = 6.23 \times 10^{-4}$

$Q > K_c$. The reaction should go to the left.

b. $Q = \dfrac{(0.0159)^2(0.0139)}{(0.121)^2} = 2.4\underline{0}0 \times 10^{-4} = 2.40 \times 10^{-4}$

$Q < K_c$. The reaction should go to the right.

c. $Q = \dfrac{(0.0134)^2(0.0181)}{(0.103)^2} = 3.0\underline{6}3 \times 10^{-4} = 3.06 \times 10^{-4}$

$Q = K_c$. The reaction should be at equilibrium.

14.39   Calculate Q, the reaction quotient, and compare it to the equilibrium constant. If Q is larger, the reaction will go to the left, and vice versa. Q is found by combining these terms:

$$Q = \frac{[CH_3OH]}{[CO][H_2]^2} = \frac{(0.020)}{(0.010)(0.010)^2} = 2\underline{0}.0\ (>10.5)$$

The reaction goes to the left.

14.41   Divide moles of substance by the volume of 5.0 L to obtain concentration. The starting concentrations are 3.0×10^{-4} M for both $[I_2]$ and $[Br_2]$. Assemble a table of starting, change, and equilibrium concentrations.

Conc. (M)	$I_2(g)$	+	$Br_2(g)$	$\rightleftharpoons$	2 IBr(g)
Starting	3.0×10^{-4}		3.0×10^{-4}		0
Change	$-x$		$-x$		$+2x$
Equilibrium	$(3.0 \times 10^{-4}) - x$		$(3.0 \times 10^{-4}) - x$		$2x$

(continued)

Substituting into the equilibrium-constant expression gives

$$K_c = 1.2 \times 10^2 = \frac{[IBr]^2}{[I_2][Br_2]} = \frac{(2x)^2}{(3.0 \times 10^{-4} - x)(3.0 \times 10^{-4} - x)}$$

Taking the square root of both sides yields

$$10.95 = \frac{(2x)}{(3.0 \times 10^{-4} - x)}$$

Rearranging and simplifying the right side gives

$$(3.0 \times 10^{-4} - x) = \frac{(2x)}{10.95} = 0.182x$$

$$x = 2.53 \times 10^{-4} \ M$$

Thus, $[I_2] = [Br_2] = 4.70 \times 10^{-5} = 4.7 \times 10^{-5} \ M$, and $[IBr] = 5.06 \times 10^{-4} = 5.1 \times 10^{-4} \ M$.

14.43 Divide moles of substance by the volume of 4.0 L to obtain concentration. The starting concentrations are 0.10 M for both $[PCl_3]$ and $[Cl_2]$. Assemble a table of starting, change, and equilibrium concentrations.

Conc. (M)	$PCl_3(g)$	+	$Cl_2(g)$	⇌	$PCl_5(g)$
Starting	0.10		0.10		0
Change	-x		-x		+x
Equilibrium	0.10 - x		0.10 - x		x

Substituting into the equilibrium expression for K_c gives

$$K_c = 49 = \frac{[PCl_5]}{[PCl_3][Cl_2]} = \frac{x}{(0.10 - x)(0.10 - x)}$$

Rearranging and solving for x yields

$$49(0.10 - x)^2 = 49(x^2 - 0.20x + 0.010) = x$$

$$49x^2 - 10.8x + 0.49 = 0 \ \text{(quadratic equation)}$$

(continued)

Using the solution to the quadratic equation gives

$$x = \frac{10.8 \pm \sqrt{(10.8)^2 - 4(49)(0.49)}}{2(49)}$$

$x = 0.157$ (impossible; reject), or $x = 0.06\underline{3}89 = 0.064\ M$ (logical)

Thus, $[PCl_3] = [Cl_2] = 0.036\ M$, and $[PCl_5] = 0.064\ M$; the vessel contains 0.18 mol PCl_3, 0.18 mol Cl_2, and 0.32 mol PCl_5.

14.45   Forward direction

14.47 a. A pressure increase has no effect because the number of moles of reactants equals that of products.

   b. A pressure increase has no effect because the number of moles of reactants equals that of products.

   c. A pressure increase causes the reaction to go to the left because the number of moles of reactants is less than that of products.

14.49   The fraction would not increase because an increase in temperature decreases the amounts of products of an exothermic reaction.

14.51   The value of $\Delta H°$ is calculated from the $\Delta H°_f$ values below each substance in the reaction:

$2NO_2(g)$	+	$7H_2(g)$	$\rightleftharpoons$	$2NH_3(g)$	+	$4H_2O(g)$
2(33.2)		7(0)		2(-45.9)		4(-241.8)

$\Delta H° = -967.2 + (-91.8) - 66.4 = -1125.4\ kJ/2\ mol\ NO_2$

The equilibrium constant will decrease with temperature because raising the temperature of an exothermic reaction will cause the reaction to go farther to the left.

▪ Solutions to General Problems

14.53 Substitute the concentrations into the equilibrium expression to calculate K_c.

$$K_c = \frac{[CH_3OH]}{[CO][H_2]^2} = \frac{(0.015)}{(0.096)(0.191)^2} = 4.\underline{2}8 = 4.3$$

14.55 After calculating the concentrations after mixing, calculate Q, the reaction quotient, and compare it with K_c.

$[N_2] = [H_2] = 2.00$ mol $\div$ 4.00 L $= 0.500$ M

$[NH_3] = 4.00$ mol $\div$ 4.00 L $= 1.00$ M

$$Q = \frac{[NH_3]^2}{[N_2][H_2]^3} = \frac{(1.00)^2}{(0.500)(0.500)^3} = 16.\underline{0}0 = 16.0$$

Because Q is greater than K_c, the reaction will go in the reverse direction (to the left) to reach equilibrium.

14.57 Assemble a table of starting, change, and equilibrium concentrations, letting $2x$ = the change in [HBr].

Conc. (M)	2HBr(g) $\rightleftharpoons$	H$_2$(g)	+	Br$_2$(g)
Starting	0.010	0		0
Change	-2x	+x		+x
Equilibrium	0.010 - 2x	x		x

$$K_c = 0.016 = \frac{[H_2][Br_2]}{[HBr]^2} = \frac{(x)(x)}{(0.010 - 2x)^2}$$

$$0.1\underline{2}6 = \frac{(x)}{(0.010 - 2x)}$$

$1.26 \times 10^{-3} - (2.52 \times 10^{-1})x = x$

$x = (1.26 \times 10^{-3}) \div 1.252 = 1.\underline{0}06 \times 10^{-3} = 1.0 \times 10^{-3}$ M

Therefore, [HBr] $= 0.008$ M, or 0.008 mol; [H$_2$] $= 0.0010$ M, or 0.0010 mol; and [Br$_2$] $= 0.0010$ M, or 0.0010 mol.

14.59   The starting concentration of $COCl_2$ = 2.00 mol $\div$ 50.00 L = 0.0400 M. Assemble a
table of starting, change, and equilibrium concentrations.

Conc. (M)	$COCl_2(g)$ $\rightleftharpoons$	$Cl_2(g)$ +	$CO(g)$
Starting	0.0400	0	0
Change	-x	+x	+x
Equilibrium	0.0400 - x	x	x

Substituting into the equilibrium expression for K_c gives

$$K_c = 8.05 \times 10^{-4} = \frac{[CO][Cl_2]}{[COCl_2]} = \frac{(x)(x)}{(0.0400 - x)}$$

Rearranging and solving for x yields

$$3.22 \times 10^{-5} - (8.05 \times 10^{-4})x - x^2 = 0$$

$$x^2 + (8.05 \times 10^{-4})x - 3.22 \times 10^{-5} = 0 \quad \text{(quadratic equation)}$$

Using the solution to the quadratic equation gives

$$x = \frac{-(8.05 \times 10^{-4}) \pm \sqrt{(8.05 \times 10^{-4})^2 - 4(1)(-3.22 \times 10^{-5})}}{2(1)}$$

$x = -6.09 \times 10^{-3}$ (impossible; reject), or $x = 5.286 \times 10^{-3}$ (logical; use)

Percent dissoc. = (change $\div$ starting) $\times$ 100% = (0.005286 $\div$ 0.0400) $\times$ 100%

$$= 13.21 = 13.2 \text{ percent}$$

14.61   The dissociation is endothermic.

14.63   For $N_2 + 3H_2 \rightleftharpoons 2NH_3$, K_p is defined in terms of pressures as

$$K_p = \frac{P_{NH_3}^2}{P_{N_2}P_{H_2}^3}$$

But by the ideal gas law, where [i] = mol/L,

$$P_i = (n_iRT)/V, \text{ or } P_i = [i]RT$$

(continued)

Substituting the right-hand equality into the K_p expression gives

$$K_p = \frac{[NH_3]^2 (RT)^2}{[N_2](RT)[H_2]^3 (RT)^3} = \frac{[NH_3]^2}{[N_2][H_2]^3}(RT)^{-2}$$

$$K_p = K_c(RT)^{-2}, \text{ or } K_c = K_p(RT)^2$$

14.65 a. The molar mass of PCl_5 is 208.22 g/mol. Thus the initial concentration of PCl_5 is

$$\frac{35.8 \text{ g } PCl_5 \times \dfrac{1 \text{ mol } PCl_5}{208.22 \text{ g } PCl_5}}{5.0 \text{ L}} = 0.03\underline{4}4 \text{ M}$$

Use the table approach, and give the starting, change, and equilibrium concentrations.

Conc. (M)	$PCl_3(g)$	+	$Cl_2(g)$	$\rightleftharpoons$	$PCl_5(g)$
Starting	0		0		0.03$\underline{4}$4
Change	x		x		$-x$
Equilibrium	x		x		0.03$\underline{4}$4 $- x$

Substituting into the equilibrium-constant expression gives

$$K_c = 4.1 = \frac{[PCl_5]}{[PCl_3][Cl_2]} = \frac{0.0344 - x}{x^2}$$

Rearranging and solving for x gives a quadratic equation.

$$4.1x^2 + x - 0.0344 = 0$$

Using the quadratic formula gives

$$x = \frac{-1 \pm \sqrt{(1)^2 - (4)(4.1)(-0.0344)}}{2(4.1)}$$

$x = 0.030\underline{6}$ (positive root)

Thus, at equilibrium, $[PCl_3] = [Cl_2] = x = 0.031 \text{ M}$. The concentration of PCl_5 is

$[PCl_5] = .03\underline{4}4 - x = 0.0344 - 0.0306 = 0.003\underline{8} = 0.004 \text{ M}$.

(continued)

b. The fraction of PCl_5 decomposed is

$$\text{fraction decomposed} = \frac{0.0306\ M}{0.0344\ M} = 0.08\underline{8}9 = 0.89$$

c. There would be a greater pressure, so less PCl_5 would decompose in order to minimize the increase in pressure.

14.67 The initial moles of $SbCl_5$ (molar mass 299.01 g/mol) are

$$65.4\ g\ SbCl_5\ \times\ \frac{1\ mol\ SbCl_5}{299.01\ g\ SbCl_5} = 0.21\underline{8}7\ mol$$

The initial pressure of $SbCl_5$ is

$$P = \frac{nRT}{V} = \frac{(0.2187\ mol)\,(0.08206\ L\bullet atm\,/\,K\bullet mol)\,(468\ K)}{5.00\ L} = 1.6\underline{8}0\ atm$$

Use the table approach, and give the starting, change, and equilibrium pressures in atm.

Press. (atm)	$SbCl_5(g)$	⇌	$SbCl_3(g)$	+	$Cl_2(g)$
Starting	1.680		0		0
Change	-x		x		x
Equilibrium	(1.680)(0.642)		(1.680)(0.358)		(1.680)(0.358)

At equilibrium, 35.8 percent of the $SbCl_5$ is decomposed, so x = (1.680)(0.358) in this table. The equilibrium-constant expression is

$$K_p = \frac{P_{SbCl_3}\bullet P_{Cl_2}}{P_{SbCl_5}} = \frac{(1.680 \times 0.358)^2}{1.680 \times 0.642} = 0.33\underline{5}4 = 0.335$$

14.69 a. The initial concentration of SO_2Cl_2 (molar mass 134.97 g/mol) is

$$\frac{8.25 \text{ g } SO_2Cl_2 \times \dfrac{1 \text{ mol } SO_2Cl_2}{134.97 \text{ g } SO_2Cl_2}}{1.00 \text{ L}} = 0.06112 \ M$$

Use the table approach, and give the starting, change, and equilibrium concentrations.

Conc. (M)	$SO_2Cl_2(g)$	$\rightleftharpoons$	$SO_2(g)$	+	$Cl_2(g)$
Starting	0.06112		0		0
Change	-x		x		x
Equilibrium	0.06112 - x		x		x

Substituting into the equilibrium-constant expression gives

$$K_c = \frac{[SO_2][Cl_2]}{[SO_2Cl_2]} = \frac{x^2}{(0.06112 - x)} = 0.045$$

Rearranging and solving for x gives a quadratic equation.

$$x^2 + 0.045x - 0.0027506 = 0$$

Using the quadratic formula gives

$$x = \frac{-(0.045) \pm \sqrt{(0.045)^2 - (4)(1)(-0.0027506)}}{2(1)}$$

$x = 0.03456$ (positive root)

The concentrations at equilibrium are $[SO_2] = [Cl_2] = x = 0.034 \ M$. For SO_2Cl_2,

$[SO_2Cl_2] = 0.06112 - x = 0.06112 - 0.03456 = 0.02655 = 0.027 \ M$.

b. The fraction of SO_2Cl_2 decomposed is

$$\text{Fraction decomposed} = \frac{0.03456 \ M}{0.06112 \ M} = 0.565 = 0.57$$

c. This will shift the equilibrium to the left and decrease the fraction of SO_2Cl_2 that has decomposed.

14.71 a. First determine the initial concentration of the dimer assuming complete reaction. The reaction can be described as 2A → D. Therefore, the initial concentration of dimer is one-half of the concentration of monomer, or 2.0×10^{-4} M. Next allow the dimer to dissociate into the monomer in equilibrium. Use the table approach, and give the starting, change, and equilibrium concentrations.

Conc. (M)	D(g)	⇌	2A(g)
Starting	2.0×10^{-4}		0
Change	$-x$		$2x$
Equilibrium	2.0×10^{-4} - x		$2x$

Substituting into the equilibrium-constant expression gives

$$K_c = \frac{[A]^2}{[D]} = \frac{(2x)^2}{(2.0 \times 10^{-4} - x)} = \frac{1}{3.2 \times 10^4} = 3.125 \times 10^{-5}$$

Rearranging and solving for x gives a quadratic equation.

$$4x^2 + (3.125 \times 10^{-5})x - (6.250 \times 10^{-9}) = 0$$

Using the quadratic formula gives

$$x = \frac{-(3.125 \times 10^{-5}) \pm \sqrt{(3.125 \times 10^{-5})^2 - (4)(4)(-6.250 \times 10^{-9})}}{2(4)}$$

$x = 3.\underline{5}8 \times 10^{-5}$ (positive root)

Thus, the concentrations at equilibrium are

$[CH_3COOH] = 2x = 2(3.\underline{5}8 \times 10^{-5}) = 7.\underline{1}6 \times 10^{-5} = 7.2 \times 10^{-5}$ M

$[Dimer] = 2.0 \times 10^{-4} - 3.\underline{5}8 \times 10^{-5} = 1.\underline{6}4 \times 10^{-4} = 1.6 \times 10^{-4}$ M

b. Some hydrogen bonding can occur that results in a more stable system. The proposed structure of the dimer is

c. An increase in the temperature would facilitate bond breaking and would decrease the amount of dimer. We could also use a Le Châtelier-type argument.

14.73 The molar mass of Br_2 is 159.82 g/mol. Thus the initial concentration of Br_2 is

$$\frac{18.22 \text{ g } Br_2 \times \dfrac{1 \text{ mol } Br_2}{159.82 \text{ g } Br_2}}{1.00 \text{ L}} = 0.11\underline{4}0 \ M$$

Use the table approach, and give the starting, change, and equilibrium concentrations.

Conc. (M)	2NO(g)	+	Br$_2$(g)	⇌	2NOBr(g)
Starting	0.112		0.11$\underline{4}$0		0
Change	-2x		-x		2x
Equilibrium	0.112 - 2x		0.11$\underline{4}$0 - x		2x (= 0.0824 M)
	= 0.02$\underline{9}$60		= 0.07$\underline{2}$80		= 0.0824

Substituting into the equilibrium-constant expression gives

$$K_c = \frac{[NOBr]^2}{[NO]^2\,[Br_2]} = \frac{(0.0824)^2}{(0.02960)^2\,(0.07280)} = 10\underline{6}.4 = 1.1 \times 10^2$$

■ Solutions to Cumulative-Skills Problems

14.75 For $Sb_2S_3(s) + 3H_2(g) \rightleftharpoons 2Sb(s) + 3H_2S(g)$ [$+3Pb^{2+} \rightarrow 3PbS(s) + 6H^+$]:

Starting M of $H_2(g)$ = 0.0100 mol ÷ 2.50 L = 0.00400 M H_2

1.029 g PbS ÷ 239.26 g PbS/mol H_2S = 4.30$\underline{0}$7 x 10^{-3} mol H_2S

4.3007 x 10^{-3} mol H_2S ÷ 2.50 L = [1.7203 x 10^{-3}] = M of H_2S

Conc. (M)	3H$_2$(g)	+	Sb$_2$S$_3$(s)	⇌	3H$_2$S(g)	+	2Sb(s)
Starting	0.00400				0		
Change	-0.0017203				+0.0017203		
Equilibrium	0.0022797				0.0017203		

Substituting into the equilibrium expression for K_c gives

$$K_c = \frac{[H_2S]^3}{[H_2]^3} = \frac{(0.0017203)^3}{(0.0022797)^3} = 0.42\underline{9}7 = 0.430$$

14.77 For $PCl_5(g) \rightleftharpoons PCl_3(g) + Cl_2(g)$:

Starting M of PCl_5 = 0.0100 mol ÷ 2.00 L = 0.00500 M

Conc. (M)	$PCl_5(g)$	$\rightleftharpoons$	$PCl_3(g)$	+	$Cl_2(g)$
Starting	0.00500		0		0
Change	-x		+x		+x
Equilibrium	0.00500 - x		x		x

Substituting into the equilibrium expression for K_c gives

$$K_c = 4.15 \times 10^{-2} = \frac{[PCl_3][Cl_2]}{[PCl_5]} = \frac{(x)(x)}{(0.00500 - x)}$$

$$x^2 + (4.15 \times 10^{-2})x - 2.075 \times 10^{-4} = 0 \text{ (quadratic)}$$

Solving the quadratic equation gives $x = -4.60 \times 10^{-2}$ (impossible) and $x = 4.51 \times 10^{-3} M$ (use).

Total M of gas = 0.00451 + 0.00451 + 0.00049 = 0.009510 M

$P = (n/V)RT$ = (0.009510 M)[0.082057 L•atm/(K•mol)](523 K)

= 0.4081 = 0.408 atm

15. ACIDS AND BASES

■ Answers to Review Questions

15.1 You can classify these acids using the information in Section 15.1. Also recall that all diatomic acids of Group VIIA halides are strong except for HF.

a. Weak   b. Weak   c. Strong

d. Strong   e. Weak   f. Weak

15.2 A Brønsted-Lowry acid is a molecule or ion that donates an H_3O^+ ion (proton donor) to a base in a proton-transfer reaction. A Brønsted-Lowry base is a molecule or ion that accepts an H_3O^+ ion (proton acceptor) from an acid in a proton-transfer reaction. An example of an acid-base equation:

$$HF(aq) + NH_3(aq) \rightarrow NH_4^+(aq) + F^-(aq)$$
$$\text{acid} \qquad \text{base} \qquad\qquad \text{acid} \qquad\qquad \text{base}$$

15.3 The conjugate acid of a base is a species that differs from the base by only one H_3O^+. Consider the base, HSO_3^-. Its conjugate acid would be H_2SO_3 but not H_2SO_4. H_2SO_4 differs from HSO_3^- by one H and one O.

15.4 You can write the equations by considering that $H_2PO_3^-$ is both a Brønsted-Lowry acid and a Brønsted-Lowry base. The $H_2PO_3^-$ acts as a Brønsted-Lowry acid when it reacts with a base such as OH^- :

$$H_2PO_3^-(aq) + OH^-(aq) \rightarrow HPO_3^{2-}(aq) + H_2O(l)$$

The $H_2PO_3^-$ acts as a Brønsted-Lowry base when it reacts with an acid such as HF:

$$H_2PO_3^-(aq) + HCl(aq) \rightarrow H_3PO_3(aq) + Cl^-(aq)$$

15.5   The Brønsted-Lowry concept enlarges on the Arrhenius concept in the following ways: (1) It expands the concept of a base to include any species that accepts protons, not just the OH⁻ ion or compounds containing the OH⁻ ion. (2) It enlarges the concepts of acids and bases to include ions as well as molecules. (3) It enables us to write acid-base reactions in nonaqueous solutions as well as in aqueous solutions, whereas the Arrhenius concept applies only to aqueous solutions. (4) It allows some species to be considered as acids or bases depending on the other reactant with which they are mixed.

15.6   According to the Lewis concept, an acid is an electron-pair acceptor, and a base is an electron-pair donor. An example is

$$Ag^+(aq) + 2(:NH_3) \rightarrow Ag(NH_3)_2^+(aq)$$

acid   base

15.7   Two factors that determine the strength of an acid are (1) the polarity of the bond to which the H atom is attached, and (2) the strength of the bond, or how tightly the proton is held by the atom to which it is bonded. An increase in the polarity of the bond makes it easier to remove the proton, increasing the strength of the acid. An increase in the strength of the bond makes it more difficult to remove the proton, decreasing the strength of the acid. The strength of the bond depends in turn on the size of the atom, so larger atoms have weaker bonds, whereas smaller atoms have stronger bonds.

15.8   The self-ionization of water is the reaction of two water molecules in which a proton is transferred from one molecule to the other to form H_3O^+ and OH⁻ ions. At 25°C, the K_w expression is $K_w = [H_3O^+][OH^-] = 1.0 \times 10^{-14}$.

15.9   The pH = -log $[H_3O^+]$ of an aqueous solution. Measure pH by using electrodes and a pH meter or by interpolating the pH from the color changes of a series of acid-base indicators.

15.10   For a neutral solution, $[H_3O^+] = [OH^-]$; thus, the $[H_3O^+]$ of a neutral solution at 37°C is the square root of K_w at 37°C:

$$[H_3O^+] = \sqrt{2.5 \times 10^{-14}} = 1.\underline{5}8 \times 10^{-7}\ M$$

$$pH = -log(1.58 \times 10^{-7}) = 6.8\underline{0}1 = 6.80$$

■ Solutions to Practice Problems

Note on significant figures: If the final answer to a solution needs to be rounded off, it is given first with one nonsignificant figure, and the last significant figure is underlined. The final answer is then rounded to the correct number of significant figures. In multiple-step problems, intermediate answers are given with at least one nonsignificant figure; however, only the final answer has been rounded off.

15.15 a. PO_4^{3-} b. HS^- c. NO_3^- d. $HAsO_4^{2-}$

15.17 a. $HBrO$ b. PH_4^+ c. $H_2PO_4^-$ d. $HSeO_3^-$

15.19 Each equation is given below with the labels for acid or base:

a. $HSO_4^-(aq) + NH_3(aq) \rightleftharpoons SO_4^{2-}(aq) + NH_4^+(aq)$

     acid   base   base   acid

The conjugate acid-base pairs are: HSO_4^-, SO_4^{2-}, and NH_4^+, NH_3.

b. $HPO_4^{2-}(aq) + NH_4^+(aq) \rightleftharpoons H_2PO_4^-(aq) + NH_3(aq)$

     base   acid   acid   base

The conjugate acid-base pairs are: $H_2PO_4^-$, HPO_4^{2-}, and NH_4^+, NH_3.

c. $Al(H_2O)_6^{3+}(aq) + H_2O(l) \rightleftharpoons Al(H_2O)_5(OH)^{2+}(aq) + H_3O^+(aq)$

     acid   base   base   acid

The conjugate acid-base pairs are: $Al(H_2O)_6^{3+}$, $Al(H_2O)_5(OH)^{2+}$, and H_3O^+, H_2O.

15.21 The reaction is

Lewis base   Lewis acid

15.23 a. Each water molecule donates a pair of electrons to iron(III), making the water molecule a Lewis base and the Fe^{3+} ion a Lewis acid.

b. The fluorine atom donates a pair of electrons to the boron atom in BF_3, making F^- a Lewis base and the BF_3 molecule a Lewis acid.

15.25 The equation is $H_2S + HOCH_2CH_2NH_2 \rightarrow HOCH_2CH_2NH_3^+ + HS^-$. The H_2S is a Lewis acid, and $HOCH_2CH_2NH_2$ is a Lewis base. The hydrogen ion from H_2S accepts a pair of electrons from the N atom in $HOCH_2CH_2NH_2$.

15.27 a. NH_4^+ is a weaker acid than H_3PO_4, so the left-hand species are favored at equilibrium.

b. HCN is a weaker acid than H_2S, so the left-hand species are favored at equilibrium.

c. H_2O is a weaker acid than HCO_3^-, so the right-hand species are favored at equilibrium.

15.29 Trichloroacetic acid is the stronger acid because, in general, the equilibrium favors the formation of the weaker acid, which is formic acid in this case.

15.31 a. HOBr is stronger because acid strength increases within a group as the size of the atom increases.

b. H_2TeO_3 is stronger because acid strength increases within a group as the size of the atom increases.

c. HI is stronger because I is more electronegative than Te. Within a period, acid strength increases as electronegativity increases.

15.33 a. $[H_3O^+] = 2\ M$

$$[OH^-] = \frac{K_w}{[H_3O^+]} = \frac{1.0 \times 10^{-14}}{2} = \underline{5}.0 \times 10^{-15} = 5 \times 10^{-15}\ M$$

b. $[OH^-] = 2 \times (0.012\ M) = 0.024\ M$

$$[H_3O^+] = \frac{K_w}{[OH^-]} = \frac{1.0 \times 10^{-14}}{0.024} = 4.\underline{1}6 \times 10^{-13} = 4.2 \times 10^{-13}\ M$$

15.35   Because the $Sr(OH)_2$ forms two OH^- per formula unit, the $[OH^-] = 2 \times 0.0050 = 0.010\ M$.

$$[H_3O^+] = \frac{K_w}{[OH^-]} = \frac{1.0 \times 10^{-14}}{0.010} = 1.\underline{0}0 \times 10^{-12} = 1.0 \times 10^{-12}\ M$$

15.37 a. $5 \times 10^{-6}\ M\ H_3O^+ > 1.0 \times 10^{-7}$, so the solution is acidic.

b.   Use K_w to determine $[H_3O^+]$

$$[H_3O^+] = \frac{K_w}{[OH^-]} = \frac{1.00 \times 10^{-14}}{5 \times 10^{-9}} = 2.0 \times 10^{-6} = 2 \times 10^{-6}\ M$$

Since $2 \times 10^{-6}\ M > 1.0 \times 10^{-7}$, the solution is acidic.

15.39 a.   pH 10.7, basic solution   b.   pH 1.9, acidic solution

c.   pH 7.0, neutral solution   d.   pH 4.3, acidic solution

15.41   Record the same number of places after the decimal point in the pH as the number of significant figures in the $[H_3O^+]$.

a. $-\log(1.0 \times 10^{-6}) = 6.0\underline{0}0 = 6.00$

b. $-\log(5.0 \times 10^{-11}) = 10.3\underline{0}1 = 10.30$

15.43 a. $pOH = -\log(5.25 \times 10^{-9}) = 8.27\underline{9}8;\ pH = 14.00 - 8.2798 = 5.7\underline{2}02 = 5.72$

b. $pOH = -\log(8.3 \times 10^{-3}) = 2.0\underline{8}09;\ pH = 14.00 - 2.0809 = 11.9\underline{1}908 = 11.92$

15.45   First convert the $[OH^-]$ to $[H_3O^+]$ using the K_w equation. Then find the pH, recording the same number of places after the decimal point in the pH as the number of significant figures in the $[H_3O^+]$.

$$[H_3O^+] = K_w \div [OH^-] = (1.0 \times 10^{-14}) \div (0.0040) = 2.\underline{5}0 \times 10^{-12}\ M$$

$$pH = -\log(2.50 \times 10^{-12}) = 11.6\underline{0}2 = 11.60$$

15.47   From the definition, pH = -log [H_3O^+] and -pH = log [H_3O^+], so enter the negative value of the pH on the calculator, and use the inverse and log keys (or 10^x) key to find the antilog of -pH. Then use the K_w equation to calculate [OH^-] from [H_3O^+].

$$log\ [H_3O^+]\ =\ -pH\ =\ -11.85$$

$$[H_3O^+]\ =\ antilog(-11.85)\ =\ 10^{-11.85}\ =\ 1.\underline{4}1 \times 10^{-12}\ M$$

$$[OH^-]\ =\ K_w \div [H_3O^+]\ =\ (1.0 \times 10^{-14}) \div (1.41 \times 10^{-12})\ =\ 7.\underline{0}7 \times 10^{-3}$$

$$=\ 7.1 \times 10^{-3}\ M$$

15.49   First calculate the molarity of the OH^- ion from the mass of NaOH. Then convert the [OH^-] to [H_3O^+] using the K_w equation. Then find the pH, recording the same number of places after the decimal point in the pH as the number of significant figures in the [H_3O^+].

$$\frac{5.80\ g\ NaOH}{1.00\ L} \quad X \quad \frac{1\ mol\ NaOH}{40.01\ g\ NaOH}\ =\ \frac{0.1450\ mol\ NaOH}{1.00\ L}\ =\ 0.145\underline{0}\ M\ OH^-$$

$$[H_3O^+]\ =\ K_w \div [OH^-]\ =\ (1.0 \times 10^{-14}) \div (0.1450)\ =\ 6.\underline{8}96 \times 10^{-14}\ M$$

$$pH\ =\ -log[H_3O^+]\ =\ -log(6.896 \times 10^{-14})\ =\ 13.1\underline{6}14\ =\ 13.16$$

15.51   Figure 15.7 shows the methyl-red indicator is yellow at pH values above about 5.5 (slightly past the midpoint of the range for methyl red). Bromthymol blue is yellow at pH values up to about 6.5 (slightly below the midpoint of the range for bromthymol blue). Therefore, the pH of the solution is between 5.5 and 6.5, and the solution is acidic.

■ Solutions to General Problems

15.53 a. BaO is a base; $BaO + H_2O \rightarrow Ba^{2+} + 2OH^-$

b. H_2S is an acid; $H_2S + H_2O \rightarrow H_3O^+ + HS^-$

c. CH_3NH_2 is a base; $CH_3NH_2 + H_2O \rightarrow CH_3NH_3^+ + OH^-$

d. SO_2 is an acid; $SO_2 + 2H_2O \rightarrow H_3O^+ + HSO_3^-$

15.55 a. $HNO_2(aq) + OH^-(aq) \rightarrow NO_2^-(aq) + H_2O(l)$

b. $HPO_4^{2-}(aq) + OH^-(aq) \rightarrow PO_4^{3-}(aq) + H_2O(l)$

c. $N_2H_5^+(aq) + OH^-(aq) \rightarrow N_2H_4(aq) + H_2O(l)$

d. $H_2AsO_4^-(aq) + OH^-(aq) \rightarrow HAsO_4^{2-}(aq) + H_2O(l)$

15.57 a. The ClO^- ion is a Brønsted base, and water is a Brønsted acid. The complete chemical equation is $ClO^-(aq) + H_2O(l) \rightleftharpoons HClO(aq) + OH^-(aq)$. The equilibrium does not favor the products because ClO^- is a weaker base than OH^-. In Lewis language, a proton from H_2O acts as a Lewis acid by accepting a pair of electrons on the oxygen of ClO^-.

$$H^+ + \left[:\overset{..}{\underset{..}{Cl}}:\overset{..}{\underset{..}{O}}: \right]^- \longrightarrow :\overset{..}{\underset{..}{Cl}}:\overset{..}{\underset{..}{O}}:H$$

b. The NH_2^- ion is a Brønsted base, and NH_4^+ is a Brønsted acid. The complete chemical equation is $NH_4^+ + NH_2^- \rightleftharpoons 2NH_3$. The equilibrium favors the products because the reactants form the solvent, a weakly ionized molecule. In Lewis language, the proton from NH_4^+ acts as a Lewis acid by accepting a pair of electrons on the nitrogen of NH_2^-.

$$\left[\begin{matrix} \overset{..}{H} \\ H:\overset{}{N}:H \\ \underset{}{H} \end{matrix} \right]^+ + \left[:\overset{..}{N}:H \\ \underset{}{H} \right]^- \longrightarrow 2\ H:\overset{..}{N}:H \\ \underset{}{H}$$

15.59 The order is $H_2S < H_2Se < HBr$. H_2Se is stronger than H_2S because, within a group, acid strength increases with increasing size of the central atom in binary acids. HBr is a strong acid, whereas the others are weak acids.

15.61 Enter the H_3O^+ concentration of 1.5×10^{-3} into the calculator, press the log key, and press the sign key to change the negative log to a positive log. This follows the negative log definition of pH. The number of decimal places of the pH should equal the significant figures in the H_3O^+.

$$pH = -\log[H_3O^+] = -\log(1.5 \times 10^{-3}) = 2.8\underline{23} = 2.82$$

15.63   Find the pOH from the pH using pH + pOH = 14.00. Then calculate the [OH⁻] from the pOH by entering the pOH into the calculator, pressing the sign key to change the positive log to a negative log, and finding the antilog. On some calculators, the antilog is found by using the inverse of the log; on other calculators, the antilog is found using the 10^x key. The number of significant figures in the [OH⁻] should equal the number of decimal places in the pOH.

$$pOH = 14.00 - 3.15 = 10.85$$

$$[OH^-] = antilog(-10.85) = 10^{-10.85} = 1.\underline{4}2 \times 10^{-11} = 1.4 \times 10^{-11} \; M$$

15.65 a. $H_2SO_4(aq) + 2NaHCO_3(aq) \rightarrow Na_2SO_4(aq) + 2CO_2(g) + 2H_2O(l)$ (molecular)

$H_3O^+(aq) + HCO_3^-(aq) \rightarrow CO_2(g) + 2H_2O(l)$ (net ionic)

b. The total moles of H_3O^+ from the H_2SO_4 is

$$mol \; H_3O^+ = \frac{0.437 \; mol \; H_2SO_4}{1 \; L} \times 0.02500 \; L \times \frac{2 \; mol \; H_3O^+}{1 \; mol \; H_2SO_4}$$

$$= 0.021\underline{8}5 \; mol \; H_3O^+$$

The moles of H_3O^+ that reacted with the NaOH are given by

$$mol \; H_3O^+ = \frac{0.108 \; mol \; NaOH}{1 \; L} \times 0.0287 \; L = 0.0031\underline{0}0 \; mol$$

The moles of $NaHCO_3$ present in the original sample are equal to the moles of H_3O^+ that reacted with the HCO_3^-, which is given by

Total moles H_3O^+ - moles H_3O^+ reacted with the NaOH = moles HCO_3^-

$0.021\underline{8}5 \; mol - 0.003823 \; mol = 0.018\underline{0}3 = 0.0180 \; mol \; NaHCO_3$

c. The mass of $NaHCO_3$ (molar mass 84.01 g/mol) present in the original sample is

$$0.01803 \; mol \; NaHCO_3 \times \frac{84.01 \; g \; NaHCO_3}{1 \; mol \; NaHCO_3} = 1.5\underline{1}4 \; g$$

Thus the percent $NaHCO_3$ in the original sample is given by

$$Percent \; NaHCO_3 = \frac{1.514 \; g}{2.500 \; g} \times 100\% = 60.\underline{5}6 = 60.6 \; percent$$

The percent KCl in the original sample is

Percent KCl = 100 - 60.56 = 39.\underline{4}4 = 39.4 percent

15.67 $HCO_3^-(aq) + H_2O(l) \rightleftharpoons H_3O^+(aq) + CO_3^{2-}(aq)$

$HCO_3^-(aq) + H_2O(l) \rightleftharpoons H_2CO_3(aq) + OH^-(aq)$

$HCO_3^-(aq) + Na^+(aq) + OH^-(aq) \rightleftharpoons Na^+(aq) + CO_3^{2-}(aq) + H_2O(l)$

$HCO_3^-(aq) + H^+(aq) + Cl^-(aq) \rightleftharpoons H_2O(l) + CO_2(g) + Cl^-(aq)$

15.69 The reaction of ammonia with water is given by

$NH_3(aq) + H_2O(l) \rightleftharpoons NH_4^+(aq) + OH^-(aq)$

The initial concentration of NH_3 (molar mass 17.03 g/mol) is

$$\text{Molarity} = \frac{4.25 \text{ g NH}_3 \times \dfrac{1 \text{ mol NH}_3}{17.03 \text{ g NH}_3}}{0.2500 \text{ L}} = 0.998\underline{2} \ M$$

Since the NH_3 is 0.42 percent reacted, the concentration of OH^- is

$[OH^-] = 0.9982 \ M \times 0.0042 = 0.0041\underline{9} \ M$

$pOH = -\log[OH^-] = -\log(0.00419) = 2.3\underline{7}8$

$pH = 14 - pOH = 14 - 2.3\underline{7}8 = 11.6\underline{2}2 = 11.62$

■ Solutions to Cumulative-Skills Problems

15.71 BF_3 acts as a Lewis acid, accepting an electron pair from NH_3:

$BF_3 + :NH_3 \rightarrow F_3B:NH_3$

The NH_3 acts as a Lewis base in donating an electron pair to BF_3. When 10.0 g of each are mixed, the BF_3 is the limiting reagent because it has the higher formula weight. The mass of $BF_3:NH_3$ formed is

$$10.0 \text{ g BF}_3 \times \frac{1 \text{ mol BF}_3}{67.81 \text{ g BF}_3} \times \frac{1 \text{ mol BF}_3:NH_3}{1 \text{ mol BF}_3} \times \frac{84.84 \text{ g BF}_3:NH_3}{1 \text{ mol BF}_3:NH_3}$$

$$= 12.\underline{5}1 = 12.5 \text{ g BF}_3:NH_3$$

16. ACID-BASE EQUILIBRIA

■ Answers to Review Questions

16.1   The equation is

$$HCN(aq) + H_2O(l) \rightleftharpoons H_3O^+(aq) + CN^-(aq)$$

The equilibrium-constant expression is

$$K_a = \frac{[H_3O^+]\,[CN^-]}{[HCN]}$$

16.2   HCN is the weakest acid. Its K_a of 4.9×10^{-10} is $< K_a$ of 1.7×10^{-5} of $HC_2H_3O_2$; $HClO_4$ is a strong acid, of course.

16.3   The degree of ionization of a weak acid decreases as the concentration of the acid added to the solution increases. Compared to low concentrations, at high concentrations, there is less water for each weak acid molecule to react with as the weak acid ionizes:

$$HA(aq) + H_2O(l) \rightleftharpoons H_3O^+(aq) + A^-(aq)$$

16.4   You can neglect x if C_a/K is ≥ 100. In this case, $C_a/K = [(0.0010\ M \div 6.8 \times 10^{-4}) = 1.47]$, which is significantly less than 100, and x cannot be neglected in the $(0.0010 - x)$ term. This says the degree of ionization is significant.

16.5 The ionization of the first H_3O^+ is

$$H_2PHO_3(aq) + H_2O(l) \rightleftharpoons H_3O^+(aq) + HPHO_3^-(aq)$$

The equilibrium-constant expression is

$$K_{a1} = \frac{[H_3O^+][HPHO_3^-]}{[H_2PHO_3]}$$

The ionization of the second H_3O^+ is

$$HPHO_3^-(aq) + H_2O(l) \rightleftharpoons H_3O^+(aq) + PHO_3^{2-}(aq)$$

The equilibrium-constant expression is

$$K_{a2} = \frac{[H_3O^+][PHO_3^{2-}]}{[HPHO_3^-]}$$

16.6 The balanced chemical equation for the ionization of aniline is

$$C_6H_5NH_2(aq) + H_2O(l) \rightleftharpoons C_6H_5NH_3^+(aq) + OH^-(aq)$$

The equilibrium-constant equation, or expression for K_b, is defined without an $[H_2O]$ term; this term is included in the value for K_b. The expression is

$$K_b = \frac{[C_6H_5NH_3^+][OH^-]}{[C_6H_5NH_2]}$$

16.7 First decide whether any of the three is a strong base or not. Because all the molecules are among the nitrogen-containing weak bases listed in Table 16.2, none is a strong base. Next, recognize the greater the $[OH^-]$, the stronger the weak base. Because $[OH^-]$ can be calculated from the square root of the product of K_b and concentration, the larger the K_b, the greater the $[OH^-]$ and the stronger the weak base. Thus CH_3NH_2 is the strongest of these three weak bases because its K_b is the largest.

16.8   The common-ion effect is the shift in an ionic equilibrium caused by the addition of a
        solute that furnishes an ion that is common to, or takes part in, the equilibrium. If the
        equilibrium involves the ionization of a weak acid, then the common ion is usually the
        anion formed by the ionization of the weak acid. If the equilibrium involves the
        ionization of a weak base, then the common ion is usually the cation formed by the
        ionization of the weak base. An example is the addition of F^- ion (as NaF) to a solution
        of the weak acid HF, which ionizes as shown below:

$$HF(aq) + H_2O(l) \rightleftharpoons H_3O^+(aq) + F^-(aq)$$

        The effect of adding F^- to this equilibrium is that it causes a shift in the equilibrium
        composition to the left. The additional F^- reacts with H_3O^+, lowering its concentration
        and raising the concentration of HF.

16.9   The addition of CH_3NH_3Cl to 0.10 M CH_3NH_2 exerts a common-ion effect that causes
        the equilibrium below to exhibit a shift in composition to the left:

$$CH_3NH_2 (aq) + H_2O(l) \rightleftharpoons CH_3NH_3^+(aq) + OH^-(aq)$$

        This shift lowers the equilibrium concentration of the OH^- ion, which increases the
        $[H_3O^+]$. An increase in $[H_3O^+]$ lowers the pH below 11.8. The shift in composition to the
        left occurs according to Le Châtelier's principle, which states that a system shifts to
        counteract any change in composition.

16.10   A buffer is most often a solution of a mixture of two substances that is able to resist pH
        changes when limited amounts of acid or base are added to it. A buffer must contain a
        weak acid and its conjugate (weak) base. Strong acids and/or bases cannot form
        effective buffers because a buffer acts by converting H_3O^+ (strong acid) to the
        unionized (weak) buffer acid and by converting OH^- (strong base) to the unionized
        (weak) buffer base. An example of a buffer pair is a mixture of H_2CO_3 and HCO_3^-, the
        principal buffer in blood.

16.11   The capacity of a buffer is the amount of acid or base with which the buffer can react
        before exhibiting a significant pH change. (A significant change in blood pH might
        mean 0.01-0.02 pH units; for other systems a significant change might mean 0.5 pH
        units.) A high-capacity buffer might be of the type discussed for Figure 16.9: one mol
        of buffer acid and one mol of buffer base. A low-capacity buffer might involve quite a
        bit less than these amounts: 0.01-0.05 mol of buffer acid and buffer base.

16.12   The pH of a weak base before titration is relatively high, around pH 10 for a 0.1 M
        solution of a typical weak base. As a strong acid titrant is added, the $[OH^-]$ decreases,
        and the pH decreases. At 50 percent neutralization, a buffer of equal amounts of buffer
        acid and base is formed. The $[OH^-]$ equals the K_b, or the pOH equals the pK_b. At the
        equivalence point, the pH is governed by the hydrolysis of the salt of the weak base
        formed and is usually in the pH 4-6 region. After the equivalence point, the pH
        decreases to a level just greater than the pH of the strong acid titrant.

■ Solutions to Practice Problems

Note on significant figures: If the final answer to a solution needs to be rounded off, it is given first with one nonsignificant figure, and the last significant figure is underlined. The final answer is then rounded to the correct number of significant figures. In multiple-step problems, intermediate answers are given with at least one nonsignificant figure; however, only the final answer has been rounded off.

16.19 a. $HBrO(aq) + H_2O(l) \rightleftharpoons H_3O^+(aq) + BrO^-(aq)$

b. $HClO_2(aq) + H_2O(l) \rightleftharpoons H_3O^+(aq) + ClO_2^-(aq)$

c. $HNO_2(aq) + H_2O(l) \rightleftharpoons H_3O^+(aq) + NO_2^-(aq)$

d. $HCN(aq) + H_2O(l) \rightleftharpoons H_3O^+(aq) + CN^-(aq)$

16.21   HAc will be used throughout as an abbreviation for acrylic acid and Ac⁻ for the acrylate ion. At the start, the H_3O^+ from the self-ionization of water is so small it is approximately zero. Once the acrylic acid solution is prepared, some of the 0.10 M HAc ionizes to H_3O^+ and Ac⁻. Let x equal the mol/L of HAc that ionize, forming x mol/L of H_3O^+ and x mol/L of Ac⁻ and leaving (0.10 - x) M HAc in solution. We can summarize the situation in tabular form:

Conc. (M)	HAc + H₂O ⇌	H₃O⁺	+ Ac⁻
Starting	0.10	0	0
Change	-x	+x	+x
Equilibrium	0.10 - x	x	x

The equilibrium-constant equation is:

$$K_a = \frac{[H_3O^+][Ac^-]}{[HAc]} = \frac{x^2}{(0.10 - x)}$$

The value of x can be obtained from the pH of the solution:

$$x = [H_3O^+] = \text{antilog}(-pH) = \text{antilog}(-2.63) = 2.\underline{3}4 \times 10^{-3} = 0.002\underline{3}4\ M$$

Note that (0.10 - x) = (0.10 - 0.00234) = 0.09766, which is significantly different from 0.10, so x cannot be ignored in the calculation. Thus we substitute for x in both the numerator and denominator to obtain the value of K_a:

$$K_a = \frac{x^2}{(0.10 - x)} = \frac{(0.00234)^2}{(0.10 - 0.00234)} = 5.\underline{6}06 \times 10^{-5} = 5.6 \times 10^{-5}$$

16.23 To solve, assemble a table of starting, change, and equilibrium concentrations. Use HBo as the symbol for boric acid and Bo⁻ as the symbol for $B(OH)_4^-$.

Conc. (M)	HBo + H₂O ⇌ H₃O⁺ + Bo⁻		
	HBo	H₃O⁺	Bo⁻
Starting	0.021	0	0
Change	-x	+x	+x
Equilibrium	0.021 - x	x	x

The value of x equals the value of the molarity of the H_3O^+ ion, which can be obtained from the equilibrium-constant expression. Substitute into the equilibrium-constant expression, and solve for x.

$$K_a = \frac{[H_3O^+][Bo^-]}{[HBo]} = \frac{(x)^2}{(0.021 - x)} = 5.9 \times 10^{-10}$$

Solve the equation for x, assuming x is much smaller than 0.021.

$$\frac{(x)^2}{(0.021 - x)} \cong \frac{(x)^2}{(0.021)} = 5.9 \times 10^{-10}$$

$$x^2 = 5.9 \times 10^{-10} \times (0.021) = 1.\underline{2}39 \times 10^{-11}$$

$$x = [H_3O^+] = 3.\underline{5}19 \times 10^{-6} \, M$$

Check to make sure the assumption that $(0.021 - x) \cong 0.021$ is valid:

$$0.021 - (3.519 \times 10^{-6}) = 0.02099, \text{ or} \cong 0.021 \text{ to two sig. figs.}$$

$$pH = -\log [H_3O^+] = -\log (3.519 \times 10^{-6}) = 5.4\underline{5}3 = 5.45$$

The degree of ionization is

$$\text{degree of ionization} = \frac{3.519 \times 10^{-6}}{(0.021)} = 0.0001\underline{6}7 = 0.00017 = 1.7 \times 10^{-4}$$

16.25   To solve, first convert the pH to $[H_3O^+]$, which also equals $[C_2H_3O_2^-]$, here symbolized as [Ac⁻]. Then assemble the usual table, and substitute into the equilibrium-constant expression to solve for $[HC_2H_3O_2]$, here symbolized as [HAc].

$$[H_3O^+] = antilog\ (-2.68) = 2.089 \times 10^{-3}\ M$$

Conc. (M)	HAc + H₂O ⇌	H₃O⁺ +	Ac⁻
Starting	x	0	0
Change	-2.089×10^{-3}	$+2.089 \times 10^{-3}$	$+2.089 \times 10^{-3}$
Equilibrium	$x - (2.089 \times 10^{-3})$	2.089×10^{-3}	2.089×10^{-3}

Write the equilibrium-constant expression in terms of chemical symbols, and then substitute the x and the $x - (2.089 \times 10^{-3})$ terms into the expression:

$$K_a = \frac{[H_3O^+][Ac^-]}{[HAc]} = \frac{(2.089 \times 10^{-3})^2}{(x - 2.089 \times 10^{-3})} = 1.7 \times 10^{-5}$$

Solve the equation for x, assuming 0.002089 is much smaller than x.

$$x = [HAc] \cong \frac{(2.089 \times 10^{-3})^2}{(1.7 \times 10^{-5})} = 0.256 = 0.26\ M$$

16.27   To solve, assemble the usual table of starting, change, and equilibrium concentrations of HF and F⁻ ions.

Conc. (M)	HF + H₂O ⇌	H₃O⁺ +	F⁻
Starting	0.045	0	0
Change	$-x$	$+x$	$+x$
Equilibrium	$0.045 - x$	x	x

Write the equilibrium-constant expression in terms of chemical symbols, and then substitute the terms x and $(0.045 - x)$:

$$K_a = \frac{[H_3O^+][F^-]}{[HF]} = \frac{(x)^2}{(0.045 - x)} = 6.8 \times 10^{-4}$$

(continued)

In this case, x cannot be ignored compared to 0.045 M. (If it is ignored, subtracting the calculated $[H_3O^+]$ from 0.045 yields a significant change.) The quadratic formula must be used. Reorganize the equilibrium-constant expression into the form $ax^2 + bx + c = 0$, and substitute for a, b, and c in the quadratic formula.

$$x^2 + (6.8 \times 10^{-4})\, x - (3.06 \times 10^{-5}) = 0$$

$$x = \frac{-6.8 \times 10^{-4} \pm \sqrt{(6.8 \times 10^{-4})^2 + 4(3.06 \times 10^{-5})}}{2}$$

$$x = \frac{-6.8 \times 10^{-4} \pm 0.01108}{2}$$

Use the positive root.

$$x = [H_3O^+] = 5.\underline{2}0 \times 10^{-3} = 5.2 \times 10^{-3}\ M$$

$$pH = -\log(5.20 \times 10^{-3}) = 2.2\underline{8}3 = 2.28$$

16.29 a. To solve, note that $K_{a1} = 1.2 \times 10^{-3} > K_{a2} = 3.9 \times 10^{-6}$, and hence the second ionization and K_{a2} can be neglected. Assemble a table of starting, change, and equilibrium concentrations. Let $H_2Ph = H_2C_8H_4O_4$ and $HPh^- = H\,C_8H_4O_4^-$.

Conc. (M)	H_2Ph + H_2O $\rightleftharpoons$	H_3O^+ +	HPh^-
Starting	0.015	0	0
Change	-x	+x	+x
Equilibrium	0.015 - x	x	x

Write the equilibrium-constant expression in terms of chemical symbols, and then substitute x and (0.015 - x):

$$K_{a1} = \frac{[H_3O^+]\,[HPh^-]}{[H_2Ph]} = \frac{(x)^2}{(0.015 - x)} = 1.2 \times 10^{-3} = 0.0012$$

In this case, x cannot be ignored in the (0.015 M - x) term. (If it is ignored, the calculated $[H_3O^+]$ when subtracted from 0.0015 M yields a significant change.) The quadratic formula must be used. Reorganize the equilibrium-constant expression into the form $ax^2 + bx + c = 0$, and substitute for a, b, and c in the quadratic formula.

$$x^2 + (0.0012)x - 1.80 \times 10^{-5} = 0$$

(continued)

$$x = \frac{-0.0012 \pm \sqrt{(0.0012)^2 - 4(-1.80 \times 10^{-5})}}{2}$$

$$x = \frac{-0.0012 \pm 0.008569}{2}$$

Use the positive root.

$$x = [H_3O^+] \cong 3.\underline{6}84 \times 10^{-3} = 0.0037 = 3.7 \times 10^{-3} \ M$$

b. Because $[HPh^-] \cong [H_3O^+]$, these terms cancel in the K_{a2} expression. This reduces to

$$[Ph^{2-}] = K_{a2} = 3.9 \times 10^{-6} \ M.$$

16.31   The equation is

$$CH_3NH_2(aq) + H_2O(l) \rightleftharpoons CH_3NH_3^+(aq) + OH^-(aq)$$

The K_b expression is

$$K_b = \frac{[CH_3NH_3^+][OH^-]}{[CH_3NH_2]}$$

16.33   To solve, convert the pH to $[OH^-]$:

$$pOH = 14.00 - pH = 14.00 - 11.34 = 2.66$$

$$[OH^-] = \text{antilog}(-2.66) = 2.\underline{1}88 \times 10^{-3} \ M$$

Using the symbol EtN for ethanolamine, assemble a table of starting, change, and equilibrium concentrations.

Conc. (M)	EtN + H₂O ⇌	HEtN⁺	+	OH⁻
Starting	0.15	0		0
Change	-x	+x		+x
Equilibrium	0.15 - (2.188 × 10⁻³)	2.188 × 10⁻³		2.188 × 10⁻³

Write the equilibrium-constant expression in terms of chemical symbols, and then substitute the terms, and solve for K_b:

$$K_b = \frac{[HEtN^+][OH^-]}{[EtN]} = \frac{(2.188 \times 10^{-3})^2}{(0.15 - 2.188 \times 10^{-3})} = 3.\underline{2}3 \times 10^{-5} = 3.2 \times 10^{-5}$$

16.35 To solve, assemble a table of starting, change, and equilibrium concentrations:

Conc. (M)	$CH_3NH_2 + H_2O \rightleftharpoons$	$CH_3NH_3^+ +$	OH^-
Starting	0.060	0	0
Change	-x	+x	+x
Equilibrium	0.060 - x	x	x

Write the equilibrium-constant expression in terms of chemical symbols, and then substitute the terms and the value of K_b:

$$K_b = \frac{[CH_3NH_3^+][OH^-]}{[CH_3NH_2]} = \frac{(x)^2}{(0.060 - x)} = 4.4 \times 10^{-4}$$

In this case, x cannot be ignored compared to 0.060 M. (If it is ignored, subtracting the calculated $[H_3O^+]$ from 0.060 yields a significant change.) The quadratic formula must be used. Reorganize the equilibrium-constant expression into the form $ax^2 + bx + c = 0$, and substitute for a, b, and c in the quadratic formula.

$$x^2 + (4.4 \times 10^{-4})x - 2.64 \times 10^{-5} = 0$$

$$x = \frac{-4.4 \times 10^{-4} \pm \sqrt{(4.4 \times 10^{-4})^2 + 4(2.64 \times 10^{-5})}}{2}$$

$$x = \frac{-4.4 \times 10^{-4} \pm 0.01028}{2}$$

Use the positive root.

$$x = [OH^-] = 4.\underline{9}2 \times 10^{-3} = 0.0049 = 4.9 \times 10^{-3} \, M$$

$$pOH = -\log(4.92 \times 10^{-3}) = 2.3\underline{0}7$$

$$pH = 14.00 - 2.307 = 11.6\underline{9}2 = 11.69$$

16.37 a. No hydrolysis occurs because the nitrate ion (NO_3^-) is the anion of a strong acid.

b. Hydrolysis occurs. Equation:

$$OCl^- + H_2O \rightleftharpoons HOCl + OH^-$$

(continued)

Equilibrium-constant expression:

$$K_b = \frac{K_w}{K_a} = \frac{[HOCl][OH^-]}{[OCl^-]}$$

c. Hydrolysis occurs. Equation:

$$NH_2NH_3^+ + H_2O \rightleftharpoons H_3O^+ + NH_2NH_2$$

Equilibrium-constant expression:

$$K_a = \frac{K_w}{K_b} = \frac{[H_3O^+][NH_2NH_2]}{[NH_2NH_3^+]}$$

d. No hydrolysis occurs because the bromide ion (Br$^-$) is the anion of a strong acid.

16.39   Acid ionization is

$$Zn(H_2O)_6^{2+}(aq) + H_2O(l) \rightleftharpoons Zn(H_2O)_5(OH)^+(aq) + H_3O^+(aq)$$

16.41 a. Ca(OCl)$_2$ is a salt of a strong base, Ca(OH)$_2$, and a weak acid, HOCl, so it would be expected to be basic.

b. K$_3$PO$_4$ is a salt of a strong base, KOH, and the anion of a weak acid, HPO$_4^{2-}$, so it would be expected to be basic.

c. NH$_4$CN is a salt of a weak base, NH$_3$, and a weak acid, HCN, so both ions hydrolyze. K_b for NH$_3$ is 1.8×10^{-5} and K_a for HCN is 4.9×10^{-10}. Since K_b is larger than K_a, the solution is expected to be basic.

d. CuSO$_4$ is a salt of a strong base, Cu(OH)$_2$, and a weak acid, HSO$_4^-$, so it would be expected to be basic.

16.43 a. The reaction is

$$NO_2^- + H_2O \rightleftharpoons HNO_2 + OH^-$$

The constant, K_b, is obtained by dividing K_w by the K_a of the conjugate acid, HNO$_2$:

$$K_b = \frac{K_w}{K_a} = \frac{1.0 \times 10^{-14}}{4.5 \times 10^{-4}} = 2.2\underline{2} \times 10^{-11} = 2.2 \times 10^{-11}$$

(continued)

b. The reaction is

$$\overset{+}{C_5H_5NH^+} + H_2O \rightleftharpoons C_5H_5N + H_3O^+$$

The constant, K_a, is obtained by dividing K_w by the K_b of the conjugate base, C_5H_5N:

$$K_a = \frac{K_w}{K_b} = \frac{1.0 \times 10^{-14}}{1.4 \times 10^{-9}} = 7.\underline{1}4 \times 10^{-6} = 7.1 \times 10^{-6}$$

16.45 Assemble the usual table, letting [Pr⁻] equal the equilibrium concentration of the propionate anion (the only ion that hydrolyzes). Then calculate the K_b of the Pr⁻ ion from the K_a of its conjugate acid, HPr. Assume x is much smaller than the 0.025 M concentration in the denominator, and solve for x in the numerator of the equilibrium-constant expression. Finally, calculate pOH from the [OH⁻] and pH from the pOH.

Conc. (M)	Pr⁻ + H₂O	⇌	HPr +	OH⁻
Starting	0.025		0	0
Change	-x		+x	+x
Equilibrium	0.025 - x		x	x

$$K_b = \frac{K_w}{K_a} = \frac{1.0 \times 10^{-14}}{1.3 \times 10^{-5}} = 7.\underline{6}9 \times 10^{-10}$$

Substitute into the equilibrium-constant equation.

$$K_b = \frac{[HPr][OH^-]}{[Pr^-]} = \frac{(x)^2}{(0.025 - x)} \cong \frac{(x)^2}{(0.025)} = 7.\underline{6}9 \times 10^{-10}$$

$$x = [OH^-] = [HPr] \cong 4.\underline{3}8 \times 10^{-6} = 4.4 \times 10^{-6} \ M$$

$$pOH = -\log[OH^-] = -\log(4.38 \times 10^{-6}) = 5.3\underline{5}8$$

$$pH = 14.00 - 5.358 = 8.6\underline{4}2 = 8.64$$

16.47 To solve, assemble a table of starting, change, and equilibrium concentrations for each part. For each part, assume x is much smaller than the 0.75 M starting concentration of HF. Then solve for x in the numerator of each equilibrium-constant expression by using the product of 6.8×10^{-4} and other terms.

a. 0.75 M Hydrofluoric acid, HF:

Conc. (M)	HF + H$_2$O	$\rightleftharpoons$	H$_3$O$^+$ + F$^-$	
Starting	0.75		0	0
Change	$-x$		$+x$	$+x$
Equilibrium	0.75 - x		x	x

$$K_a = \frac{[H_3O^+][F^-]}{[HF]} = \frac{(x)^2}{(0.75 - x)} \cong \frac{(x)^2}{(0.75)} = 6.8 \times 10^{-4}$$

$$x^2 = 6.8 \times 10^{-4} \times (0.75)$$

$$x = [H_3O^+] = 0.02258\ M$$

Check to see if the assumption is valid.

$$0.75 - (0.02258) = 0.727 = 0.73$$

This is a borderline case; the quadratic equation gives $[H_3O^+] = 0.02224\ M$, not much different. The degree of ionization is

$$\text{Degree of ionization} = \frac{0.02258}{0.75} = 0.0301 = 0.030$$

b. 0.75 M HF with 0.12 M HCl:

Conc. (M)	HF + H$_2$O	$\rightleftharpoons$	H$_3$O$^+$ + F$^-$	
Starting	0.75		0.12	0
Change	$-x$		$+x$	$+x$
Equilibrium	0.75 - x		0.12 + x	x

Assuming x is negligible compared to 0.12 and to 0.75, substitute into the equilibrium-constant expression 0.12 for $[H_3O^+]$ from 0.12 M HCl and 0.75 from the HF:

(continued)

$$K_a = \frac{[H_3O^+][F^-]}{[HF]} = \frac{(0.12 + x)(x)}{(0.75 - x)} \cong \frac{(0.12)(x)}{(0.75)} = 6.8 \times 10^{-4}$$

$$x = [F^-] = \frac{(0.75)(6.8 \times 10^{-4})}{(0.12)} \cong 4.\underline{2}50 \times 10^{-3}\ M$$

Check to see if the assumptions are valid:

$$0.75 - (4.250 \times 10^{-3}) = 0.7\underline{4}57 = 0.75$$

$$0.12 + (4.250 \times 10^{-3}) = 0.1\underline{2}42 = 0.12$$

$$\text{Degree of ionization} = \frac{4.25 \times 10^{-3}}{0.75} = 0.005\underline{6}6 = 0.0057$$

16.49 Assemble the usual table, using a starting NO_2^- of 0.10 M, from 0.10 M KNO_2, and a starting HNO_2 of 0.15 M. Assume x is negligible compared to 0.10 M and 0.15 M, and solve for the x in the numerator.

Conc. (M)	HNO_2 + H_2O	$\rightleftharpoons$	H_3O^+ +	NO_2^-
Starting	0.15		0	0.10
Change	-x		+x	+x
Equilibrium	0.15 - x		x	0.10 + x

$$K_a = \frac{[H_3O^+][NO_2^-]}{[HNO_2]} = \frac{(0.10 + x)(x)}{(0.15 - x)} \cong \frac{(0.10)(x)}{(0.15)} = 4.5 \times 10^{-4}$$

$$x = [H_3O^+] = 6.\underline{7}5 \times 10^{-4}\ M$$

$$pH = -\log[H_3O^+] = -\log(6.75 \times 10^{-4}) = 3.1\underline{7}0 = 3.17$$

16.51 Find the mol/L of HF and the mol/L of F^-, and assemble the usual table. Substitute the equilibrium concentrations into the equilibrium-constant expression; then assume x is negligible compared to the starting concentrations of both HF and F^-. Solve for x in the numerator of the equilibrium-constant expression, and calculate the pH from this value.

Total volume = 0.045 L + 0.035 L = 0.080 L

(0.10 mol HF/L) x 0.035 L = 0.0035 mol HF (÷ 0.080 L total volume = 0.04375 M)

(0.20 mol F^-/L) x 0.045 L = 0.0090 mol F^- (÷ 0.080 L total volume = 0.1125 M)

(continued)

Now substitute these starting concentrations into the usual table:

Conc. (M)	HF + H_2O $\rightleftharpoons$ H_3O^+ + F^-		
Starting	0.04375	0	0.1125
Change	-x	+x	+x
Equilibrium	0.04375 - x	x	0.1125 + x

$$K_a = \frac{[H_3O^+][F^-]}{[HF]} = \frac{(x)(0.1125 + x)}{(0.04375 - x)} \cong \frac{(x)(0.1125)}{(0.04375)} = 6.8 \times 10^{-4}$$

$x = [H_3O^+] = 2.\underline{6}4 \times 10^{-4} M$

$pH = -\log [H_3O^+] = -\log (2.64 \times 10^{-4}) = 3.5\underline{7}7 = 3.58$

16.53   First use the 0.10 M NH_3 and 0.10 M NH_4^+ to calculate the $[OH^-]$ and pH before HCl is added. Assemble a table of starting, change, and equilibrium concentrations. Assume x is negligible compared to 0.10 M, and substitute the approximate concentrations into the equilibrium-constant expression.

Conc. (M)	NH_3 + H_2O $\rightleftharpoons$ NH_4^+ + OH^-		
Starting	0.10	0.10	0
Change	-x	+x	+x
Equilibrium	0.10 - x	0.10 + x	x

$$K_b = \frac{[NH_4^+][OH^-]}{[NH_3]} = \frac{(0.10 + x)(x)}{(0.10 - x)} \cong \frac{(0.10)(x)}{(0.10)} = 1.8 \times 10^{-5}$$

$x = [OH^-] = 1\,8 \times 10^{-5} M$

$pOH = -\log [OH^-] = -\log (1.8 \times 10^{-5}) = 4.7\underline{4}4$

$pH = 14.00 - pOH = 14.00 - 4.744 = 9.2\underline{5}52 = 9.26$ (before HCl added)

Now calculate the pH after the 0.012 L (12 mL) of 0.20 M HCl is added by noting that the H_3O^+ ion reacts with the NH_3 to form additional NH_4^+. Calculate the stoichiometric amount of HCl; then subtract the moles of HCl from the moles of NH_3. Add the resulting moles of NH_4^+ to the 0.0125 starting moles of NH_4^+ in the 0.125 L of buffer.

(0.20 mol HCl/L) x 0.012 L = 0.0024 mol HCl (reacts with 0.0024 mol NH_3)

Mol NH_3 left = (0.01\underline{2}5 - 0.0024) mol = 0.01\underline{0}1 mol

Mol NH_4^+ present = (0.0125 + 0.0024) mol = 0.0149 mol

(continued)

The concentrations of NH_3 and NH_4^+ are

$$[NH_3] = \frac{0.0101 \text{ mol } NH_3}{0.137 \text{ L}} = 0.0737 \text{ } M$$

$$[NH_4^+] = \frac{0.0149 \text{ mol } NH_3}{0.137 \text{ L}} = 0.108 \text{ } M$$

Now account for the ionization of NH_3 to NH_4^+ and OH^- at equilibrium by assembling the usual table. Assume x is negligible compared to 0.0737 M and 0.108 M, and solve the equilibrium-constant expression for x in the numerator. Calculate the pH from x, the $[OH^-]$.

Conc. (M)	$NH_3 + H_2O$	$\rightleftharpoons$	NH_4^+	+	OH^-
Starting	0.0737		0.108		0
Change	-x		+x		+x
Equilibrium	0.0737 - x		0.108 + x		x

$$K_b = \frac{[NH_4^+][OH^-]}{[NH_3]} = \frac{(0.108 + x)(x)}{(0.0737 - x)} \cong \frac{(0.108)(x)}{(0.0737)} = 1.8 \times 10^{-5}$$

$$x = [OH^-] \cong 1.22 \times 10^{-5} \text{ } M$$

$$pOH = -\log [OH^-] = -\log (1.22 \times 10^{-5}) = 4.913$$

$$pH = 14.00 - pOH = 14.00 - 4.913 = 9.086 = 9.09 \text{ (after HCl added)}$$

16.55 Symbolize acetic acid as HOAc and sodium acetate as Na^+OAc^-. Use the Henderson-Hasselbalch equation to find the log of $[OAc^-]/[HOAc]$. Then solve for $[OAc^-]$ and for moles of NaOAc in the 2.0 L of solution.

$$pH = -\log K_a + \log \frac{[OAc^-]}{[HOAc]} = 5.00$$

$$5.00 = -\log (1.7 \times 10^{-5}) + \log \frac{[OAc^-]}{(0.10 \text{ } M)}$$

$$5.00 = 4.770 + \log [OAc^-] - \log (0.10)$$

$$\log [OAc^-] = 5.00 - 4.770 - 1.00 = -0.770$$

$$[OAc^-] = 0.1698 \text{ } M$$

Mol NaOAc = 0.1698 mol/L $\times$ 2.0 L = 0.3396 = 0.34 mol

16.57 All the OH⁻ (from the NaOH) reacts with the H₃O⁺ from HCl. Calculate the stoichiometric amounts of OH⁻ and H₃O⁺, and subtract the mol of OH⁻ from the mol of H₃O⁺. Then divide the remaining H₃O⁺ by the total volume of 0.015 L + 0.025 L, or 0.040 L, to find the [H₃O⁺]. Then calculate the pH.

Mol H_3O^+ = (0.10 mol HCl/L) x 0.025 L HCl = 0.0025 mol H_3O^+

Mol OH^- = (0.10 mol NaOH/L) x 0.015 L NaOH = 0.0015 mol OH^-

Mol H_3O^+ left = (0.0025 - 0.0015) mol H_3O^+ = 0.0010 mol H_3O^+

$[H_3O^+]$ = 0.0010 mol H_3O^+ ÷ 0.040 L total volume = 0.0250 M

pH = - log $[H_3O^+]$ = - log (0.0250) = 1.602 = 1.60

16.59 Use EtN to symbolize ethylamine and EtNH⁺ to symbolize the ethylammonium cation. At the equivalence point, equal molar amounts of EtN and HCl react to form a solution of EtNHCl. Start by calculating the moles of EtN. Use this to calculate the volume of HCl needed to neutralize all of the EtN (and use the moles of EtN as the moles of EtNH⁺ formed at the equivalence point). Add the volume of HCl to the original 0.032 L to find the total volume of solution.

(0.087 mol EtN/L) x 0.032 L = 0.00278 mol EtN

Volume HCl = 0.00278 mol HCl ÷ (0.15 mol HCl/L) = 0.0185 L

Total volume = 0.0185 L + 0.032 L EtN soln = 0.0505 L

$[EtNH^+]$ = (0.00278 mol EtNH⁺ from EtN) ÷ 0.0505 L = 0.0550 M

Because the EtNH⁺ hydrolyzes to H₃O⁺ and EtN, use this to calculate the [H₃O⁺]. Start by calculating the K_a constant of EtNH⁺ from the K_b of its conjugate base, EtN. Then assemble the usual table of concentrations, assume x is negligible, and calculate [H₃O⁺] and pH.

$$K_a = \frac{K_w}{K_b} = \frac{1.0 \times 10^{-14}}{4.7 \times 10^{-4}} = 2.13 \times 10^{-11}$$

Conc. (M)	$EtNH^+ + H_2O \rightleftharpoons$	EtN +	H_3O^+
Starting	0.0550	0	0
Change	-x	+x	+x
Equilibrium	0.0550 - x	x	x

(continued)

$$K_a = \frac{[EtN][H_3O^+]}{[EtNH^+]} = \frac{(x)^2}{(0.0550 - x)} \cong \frac{(x)^2}{(0.0550)} = 2.13 \times 10^{-11}$$

$$x = [H_3O^+] = 1.\underline{0}8 \times 10^{-6}\ M$$

$$pH = -\log[H_3O^+] = -\log(1.08 \times 10^{-6}) = 5.9\underline{6}56 = 5.97$$

16.61 Calculate the stoichiometric amounts of NH_3 and HCl, which form NH_4^+. Then divide NH_3 and NH_4^+ by the total volume of 0.500 L + 0.200 L = 0.700 L to find the starting concentrations. Calculate the $[OH^-]$, the pOH, and the pH.

Mol NH_3 = (0.10 mol NH_3/L) x 0.500 L = 0.0500 mol NH_3

Mol HCl = (0.15 mol HCl/L) x 0.200 L = 0.0300 mol HCl (0.0300 mol NH_4^+)

Mol NH_3 left = 0.0500 mol - 0.0300 mol HCl = 0.0200 mol NH_3

0.02\underline{0}0 mol NH_3 ÷ 0.700 L = 0.0286 M NH_3

0.03\underline{0}0 mol NH_4^+ ÷ 0.700 L = 0.0429 M NH_4^+

Conc. (M)	$NH_3 + H_2O \rightleftharpoons$	NH_4^+ +	OH^-
Starting	0.0286	0.0429	0
Change	-x	+x	+x
Equilibrium	0.0286 - x	0.0429 + x	x

$$K_b = \frac{[NH_4^+][OH^-]}{[NH_3]} = \frac{(0.0429 + x)(x)}{(0.0286 - x)} \cong \frac{(0.0429)(x)}{(0.0286)} = 1.8 \times 10^{-5}$$

$$x = [OH^-] = 1.\underline{2}0 \times 10^{-5}\ M$$

$$pOH = -\log[OH^-] = -\log(1.20 \times 10^{-5}) = 4.9\underline{2}0$$

$$pH = 14.00 - pOH = 14.00 - 4.920 = 9.0\underline{7}9 = 9.08$$

■ Solutions to General Problems

16.63 For the base ionization (hydrolysis) of CN^- to $HCN + OH^-$, the base-ionization constant is

$$K_b = \frac{K_w}{K_a} = \frac{1.0 \times 10^{-14}}{4.9 \times 10^{-10}} = 2.0\underline{4} \times 10^{-5} = 2.0 \times 10^{-5}$$

For the base ionization (hydrolysis) of CO_3^{2-} to $HCO_3^- + OH^-$, the base-ionization constant is calculated from the ionization constant (K_{a2}) of HCO_3^-, the conjugate acid of CO_3^{2-}.

$$K_b = \frac{K_w}{K_a} = \frac{1.0 \times 10^{-14}}{4.8 \times 10^{-11}} = 2.0\underline{8} \times 10^{-4} = 2.1 \times 10^{-4}$$

Because the constant of CO_3^{2-} is larger, it is the stronger base.

16.65 Assume Al^{3+} is $Al(H_2O)_6^{3+}$. Assemble the usual table to calculate the $[H_3O^+]$ and pH. Use the usual equilibrium-constant expression.

Conc. (M)	$Al(H_2O)_6^{3+} + H_2O \rightleftharpoons$	$H_3O^+ +$	$Al(H_2O)_5(OH)^{2+}$
Starting	0.15	0	0
Change	-x	+x	+x
Equilibrium	0.15 - x	x	x

$$K_a = \frac{[Al(H_2O)_5(OH)^{2+}][H_3O^+]}{[Al(H_2O)_6^{3+}]} = \frac{(x)^2}{(0.15 - x)} \cong \frac{(x)^2}{(0.15)} = 1.4 \times 10^{-5}$$

$$x = [H_3O^+] = 1.4\underline{4}9 \times 10^{-3} \ M$$

$$pH = -\log [H_3O^+] = -\log (1.449 \times 10^{-3}) = 2.8\underline{3}8 = 2.84$$

16.67 Use the Henderson-Hasselbalch equation where $[H_2CO_3]$ = the buffer acid, $[HCO_3^-]$ = the buffer base, and K_{a1} of carbonic acid is the ionization constant.

$$pH = -\log K_a + \log \frac{[HCO_3^-]}{[H_2CO_3]} = 7.40$$

$$7.40 = -\log (4.3 \times 10^{-7}) + \log \frac{[HCO_3^-]}{[H_2CO_3]}$$

(continued)

$$\log \frac{[HCO_3^-]}{[H_2CO_3]} = 7.40 - 6.366 = 1.0\underline{3}4$$

$$\frac{[HCO_3^-]}{[H_2CO_3]} = \frac{10.81}{1} = \frac{11}{1}$$

16.69 All the OH⁻ (from the NaOH) reacts with the H_3O^+ from HCl. Calculate the stoichiometric amounts of OH⁻ and H_3O^+, and subtract the mol of OH⁻ from the mol of H_3O^+. Then divide the remaining H_3O^+ by the total volume of 0.456 L + 0.285 L, or 0.741 L, to find the $[H_3O^+]$. Then calculate the pH.

Mol H_3O^+ = (0.10 mol HCl/L) x 0.456 L HCl = 0.0456 mol H_3O^+

Mol OH⁻ = (0.15 mol NaOH/L) x 0.285 L NaOH = 0.0428 mol OH⁻

Mol H_3O^+ left = (0.0456 - 0.0428) mol H_3O^+ = 0.0028 mol H_3O^+

$[H_3O^+]$ = 0.0028 mol H_3O^+ ÷ 0.741 L total volume = 0.00\underline{3}8 M

pH = - log $[H_3O^+]$ = - log (0.0038) = 2.\underline{4}2 = 2.4

16.71 Use BzN to symbolize benzylamine and BzNH⁺ to symbolize the benzylammonium cation. At the equivalence point, equal molar amounts of BzN and HCl react to form a solution of BzNHCl. Start by calculating the moles of BzN. Use this number to calculate the volume of HCl needed to neutralize all the BzN (and use the moles of BzN as the moles of BzNH⁺ formed at the equivalence point). Add the volume of HCl to the original 0.025 L to find the total volume of solution.

(0.025 mol BzN/L) x 0.065 L = 0.00162 mol BzN

Volume HCl = 0.00162 mol HCl ÷ (0.050 mol HCl/L) = 0.0324 L

Total volume = 0.025 L + 0.0324 L = 0.05\underline{7}4 L

$[BzNH^+]$ = (0.00162 mol BzNH⁺ from BzN) ÷ 0.0574 L = 0.02\underline{8}2 M

Because the BzNH⁺ hydrolyzes to H_3O^+ and BzN, use this to calculate the $[H_3O^+]$. Start by calculating the K_a constant of BzNH⁺ from the K_b of its conjugate base, BzN. Then assemble the usual table of concentrations, assume x is negligible, and calculate $[H_3O^+]$ and pH.

$$K_a = \frac{K_w}{K_b} = \frac{1.0 \times 10^{-14}}{4.7 \times 10^{-10}} = 2.\underline{1}3 \times 10^{-5}$$

(continued)

Conc. (M)	$BzNH^+ + H_2O$	$\rightleftharpoons$	H_3O^+	+	BzN
Starting	0.0282		0		0
Change	-x		+x		+x
Equilibrium	0.0282 - x		x		x

Substitute into the equilibrium-constant equation.

$$K_a = \frac{[H_3O^+][BzN]}{[BzNH^+]} = \frac{(x)^2}{(0.0282 - x)} \cong \frac{(x)^2}{(0.0282)} = 2.13 \times 10^{-5}$$

$$x = [H_3O^+] = 7.\underline{7}502 \times 10^{-4} \ M$$

$$pH = -\log[H_3O^+] = -\log(7.7502 \times 10^{-4}) = 3.1\underline{1}06 = 3.11$$

16.73 a. From the pH, calculate the H_3O^+ ion concentration:

$$[H_3O^+] = 10^{-pH} = 10^{-5.82} = 1.\underline{5}1 \times 10^{-6} \ M$$

Use the table approach, giving the starting, change, and equilibrium concentrations.

Conc. (M)	$CH_3NH_3^+ + H_2O$	$\rightleftharpoons$	CH_3NH_2	+	H_3O^+
Starting	0.10		0		0
Change	-x		+x		+x
Equilibrium	0.10 - x		x		x

From the hydronium ion concentration, $x = 1.51 \times 10^{-6}$. Substituting into the equilibrium-constant expression gives

$$K_a = \frac{[CH_3NH_2][H_3O^+]}{[CH_3NH_3^+]} = \frac{(x)^2}{(0.10 - x)} \cong \frac{(1.51 \times 10^{-6})^2}{(0.10)} = 2.2\underline{8} \times 10^{-11}$$

$$= 2.3 \times 10^{-11}$$

b. Now, use K_w to calculate the value of K_b:

$$K_b = \frac{K_w}{K_a} = \frac{1.0 \times 10^{-14}}{2.28 \times 10^{-11}} = 4.\underline{3}8 \times 10^{-4} = 4.4 \times 10^{-4}$$

(continued)

c. The equilibrium concentration of $CH_3NH_3^+$ is approximately 0.450 mol/1.00 L = 0.450 M. Use $[CH_3NH_2] \cong 0.250$ M. For K_a, we get

$$K_a = \frac{[CH_3NH_2][H_3O^+]}{[CH_3NH_3^+]} \cong \frac{(0.250)\,[H_3O^+]}{(0.450)} = 2.\underline{2}8 \times 10^{-11}$$

Solving for $[H_3O^+]$ gives

$$[H_3O^+] \cong \frac{(2.28 \times 10^{-11})(0.450)}{(0.250)} = 4.\underline{1}0 \times 10^{-11} \; M$$

Thus, the pH is

$$pH = -\log(4.10 \times 10^{-11}) = 10.3\underline{8}7 = 10.39$$

16.75 a. True. Weak acids have small K_a values, so most of the solute is present as undissociated molecules.

b. True. Weak acids have small K_a values, so most of the solute is present as the molecule.

c. False. The hydroxide concentration equals the hydronium concentration only in neutral solutions.

d. False. If HA were a strong acid, the pH would be equal to two.

e. False. The H_3O^+ would be 0.010 if HA were a strong acid.

f. True. For every HA molecule that dissociates, one H_3O^+ is generated along with one A^-.

16.77 a. Initial (zero percent) pH: Use the table approach, and give the starting, change, and equilibrium concentrations.

Conc. (M)	$NH_3 + H_2O \rightleftharpoons$	$NH_4^+ +$	OH^-
Starting	0.10	0	0
Change	$-x$	$+x$	$+x$
Equilibrium	$0.10 - x$	x	x

(continued)

Substituting into the equilibrium-constant expression gives

$$K_b = \frac{[NH_4^+][OH^-]}{[OH^-]} = \frac{(x)^2}{(0.10 - x)} \cong \frac{(x)^2}{(0.10)} = 1.8 \times 10^{-5}$$

Rearranging and solving for x gives

$$x = [OH^-] = \sqrt{(0.10)(1.8 \times 10^{-5})} = 1.\underline{3}4 \times 10^{-3} \ M$$

$$pOH = -\log(1.34 \times 10^{-3}) = 2.8\underline{7}3$$

$$pH = 14 - pOH = 14 - 2.873 = 11.1\underline{2}7 = 11.13$$

30-percent titration point: Express the concentrations as percents for convenience.

$[NH_3] \approx 70\%$ and $[NH_4^+] \approx 30\%$

Plug into the equilibrium-constant expression.

$$K_b = \frac{[NH_4^+][OH^-]}{[NH_3]} = \frac{(30\%)(x)}{(70\%)} = 1.8 \times 10^{-5}$$

Solving for x gives

$$x = [OH^-] = 1.8 \times 10^{-5} \times \frac{70\%}{30\%} = 4.\underline{2}0 \times 10^{-5} \ M$$

$$pOH = -\log(4.20 \times 10^{-5}) = 4.3\underline{7}7$$

$$pH = 14 - pOH = 14 - 4.377 = 9.6\underline{2}3 = 9.62$$

50-percent titration point:

$[NH_4^+] = [NH_3]$

$$K_b = [OH^-] = 1.8 \times 10^{-5} \ M$$

$$pOH = -\log(1.8 \times 10^{-5}) = 4.7\underline{4}4$$

$$pH = 14 - pOH = 14 - 4.744 = 9.2\underline{5}6 = 9.26$$

(continued)

100-percent titration point:

The NH_4Cl that is produced has undergone a two-fold dilution.

$[NH_4^+] = 0.0500\ M$

Use the table approach, and give the starting, change, and equilibrium concentrations.

Conc. (M)	$NH_4^+ + H_2O \rightleftharpoons$	$NH_3 +$	H_3O^+
Starting	0.0500	0	0
Change	-x	+x	+x
Equilibrium	0.0500 - x	x	x

Now use K_w to calculate the value of K_a:

$$K_a = \frac{K_w}{K_b} = \frac{1.0 \times 10^{-14}}{1.8 \times 10^{-5}} = 5.\underline{5}5 \times 10^{-10}$$

Plug into the equilibrium-constant expression.

$$K_a = \frac{[NH_3][H_3O^+]}{[NH_4^+]} = \frac{(x)^2}{(0.0500 - x)} \cong \frac{(x)^2}{(0.0500)} = 5.55 \times 10^{-10}$$

Solving for x gives

$$x = [H_3O^+] = \sqrt{0.0500 \times \frac{1.0 \times 10^{-14}}{1.8 \times 10^{-5}}} = 5.\underline{2}7 \times 10^{-6}\ M$$

$$pH = -\log(5.27 \times 10^{-6}) = 5.2\underline{7}8 = 5.28$$

b. The solution is acidic because the NH_4^+ ion reacts with water to produce acid. Ammonium chloride is the salt of a weak base and a strong acid.

16.79 a. Select the conjugate pair that has a pK_a value closest to a pH of 2.88.

$pK_a\ (H_2C_2O_4) = 1.25$

$pK_a\ (H_3PO_4) = 2.16$

$pK_a\ (HCOOH) = 3.77$

Therefore, the best pair is H_3PO_4 and $H_2PO_4^-$.

(continued)

b. Using the Henderson-Hasselbalch equation,

$$pH = pK_a + \log \frac{[A^-]}{[HA]}$$

$$2.88 = 2.16 + \log \frac{[A^-]}{[HA]}$$

$$\log \frac{[A^-]}{[HA]} = 0.72$$

$$\frac{[A^-]}{[HA]} = 10^{0.72} = 5.\underline{2}4$$

Therefore, 5.24 times more conjugate base is needed than acid. Since the starting concentrations of H_3PO_4 (HA) and $H_2PO_4^-$ (A$^-$) are the same, the volume of A$^-$ needed is 5.24 times the volume of HA needed. For 50 mL of buffer, this is

50 mL = vol HA + vol A$^-$ = vol HA + 5.24 vol HA = 6.24 vol HA

Therefore, the volume of 0.10 M H_3PO_4 required is

$$\text{vol } H_3PO_4 = \frac{50 \text{ mL}}{6.24} = 8.\underline{0}1 \text{ mL} = 8 \text{ mL}$$

The volume of 0.10 M $H_2PO_4^-$ required is

50 mL - 8.01 mL = 4\underline{2}.0 mL = 42 mL

16.81 a. At the equivalence point, moles of base equal moles of acid. Therefore,

$$M_{NH_2OH} \times 25.0 \text{ mL} = 0.150 \ M \times 35.8 \text{ mL}$$

$$M_{NH_2OH} = \frac{0.150 \ M \times 35.8 \text{ mL}}{25.0 \text{ mL}} = 0.21\underline{4}8 = 0.215$$

b. Concentration of NH_3OH^+ at the equivalence point is

$$[NH_3OH^+] = \frac{(25.0 \text{ mL})(0.215 \ M)}{25.0 \text{ mL} + 35.8 \text{ mL}} = 0.088\underline{4}0 \ M$$

$$K_a = \frac{K_w}{K_b} = \frac{1.0 \times 10^{-14}}{1.1 \times 10^{-8}} = 9.\underline{0}9 \times 10^{-7}$$

(continued)

Use the table approach, and give the starting, change, and equilibrium concentrations.

Conc. (M)	$NH_3OH^+ + H_2O \rightleftharpoons NH_2OH + H_3O^+$		
Starting	0.08840	0	0
Change	-x	+x	+x
Equilibrium	0.08840 - x	x	x

Substituting into the equilibrium-constant expression gives

$$K_a = \frac{[NH_2OH][H_3O^+]}{[NH_3OH^+]} = \frac{(x)^2}{(0.08840 - x)} \cong \frac{(x)^2}{(0.08840)} = 9.09 \times 10^{-7}$$

Rearranging and solving for x gives

$$x = [H_3O^+] = \sqrt{(0.08840)(9.09 \times 10^{-7})} = 2.83 \times 10^{-4} \ M$$

$$pH = -\log(2.84 \times 10^{-4}) = 3.547 = 3.55$$

c. You need an indicator that changes pH around pH of 3 to 4. Therefore, the appropriate indicator is bromophenol blue. Select an indicator that changes color around the equivalence point.

16.83 a. $H_2A + H_2O \rightleftharpoons H_3O^+ + HA^- \qquad K_{a_1}$

$HA^- + H_2O \rightleftharpoons H_3O^+ + A^{2-} \qquad K_{a_2}$

$H_2A + 2 H_2O \rightleftharpoons 2 H_3O^+ + A^{2-}, \qquad K = K_{a_1} \times K_{a_2}$

b. $H_2A \gg H_3O^+ = HA^- \gg A^{2-}$

c. $H_2A + H_2O \rightleftharpoons H_3O^+ + HA^- \qquad K_{a_1} = 1.0 \times 10^{-3}$

    $\quad 0.0250 - x \qquad\qquad\quad x \qquad x$

$$K_{a_1} = \frac{[HA^-][H_3O^+]}{[H_2A]} = \frac{(x)^2}{(0.0250 - x)} = 1.0 \times 10^{-3}$$

(continued)

Rearranging into a quadratic equation, and solving for x gives

$$x^2 + (1.0 \times 10^{-3})x + (-2.50 \times 10^{-5}) = 0$$

$$x = \frac{-(1.0 \times 10^{-3}) \pm \sqrt{(1.0 \times 10^{-3})^2 - (4)(1)(-2.50 \times 10^{-5})}}{2}$$

$$x = [H_3O^+] = [HA^-] = 4.\underline{5}2 \times 10^{-3} = 4.5 \times 10^{-3}\ M$$

$$pH = -\log(4.\underline{5}2 \times 10^{-3}) = 2.3\underline{4}4 = 2.34$$

$$[H_2A] = 0.0250 - x = 0.0250 - 4.52 \times 10^{-3} = 0.020\underline{4}8 = 0.0205\ M$$

d. $HA^- + H_2O \rightleftharpoons H_3O^+ + A^{2-}$ $\quad K_{a_2} = 4.6 \times 10^{-5}$

$\quad 4.52 \times 10^{-3} - y \qquad\qquad 4.52 \times 10^{-3} + y \quad y$

Assume y is small compared to $4.52 \times 10^{-3}\ M$. Substitute into the rearranged equilibrium-constant equation to get

$$[A^{2-}] = \frac{[HA^-]K_{a_2}}{[H_3O^+]} = \frac{(4.52 \times 10^{-3})(4.6 \times 10^{-5})}{(4.52 \times 10^{-3})} = 4.6 \times 10^{-5}\ M$$

■ Solutions to Cumulative-Skills Problems

16.85 Use the pH to calculate $[H_3O^+]$, and then use the K_a of 1.7×10^{-5} and the K_a expression to calculate the molarity of acetic acid (HAc), assuming ionization is negligible. Convert molarity to mass percentage using the formula weight of 60.05 g/mol of HAc.

$$[H_3O^+] = \text{antilog}(-2.45) = 3.\underline{5}48 \times 10^{-3}\ M$$

Write the equilibrium-constant expression in terms of chemical symbols, and then substitute the x and the $(0.003548\ M)$ terms into the expression:

$$K_a = \frac{[H_3O^+][Ac^-]}{[HAc]} \cong \frac{(0.003548)^2}{(x)}$$

Solve the equation for x, assuming 0.003548 is much smaller than x.

(continued)

$$x = (0.003548)^2 \div (1.7 \times 10^{-5}) \cong [HAc] \cong 0.7\underline{4}04 \; M$$

$$(0.7404 \; mol \; HAc/L) \times (60.05 \; g/mol) \times (1 \; L/1090 \; g)(100\%)$$

$$= 4.\underline{0}78 = 4.1 \; percent \; HAc$$

16.87   The $[H_3O^+] =$ -antilog (-4.35) = $4.\underline{4}6 \times 10^{-5} \; M$. Note that 0.465 L of 0.0941 M NaOH will produce 0.043756 mol of acetate ion, Ac⁻. Rearranging the K_a expression for acetic acid (HAc) and Ac⁻ and canceling the volume in the mol/L of each, you obtain:

$$\frac{[HAc]}{[Ac^-]} = \frac{[H^+]}{K_a} = \frac{4.46 \times 10^{-5}}{1.7 \times 10^{-5}} = \frac{2.63}{1.00} \cong \frac{x \; mol \; HAc}{0.043756 \; mol \; Ac^-}$$

$$x = 0.1\underline{1}49 \; mol \; HAc$$

Total mol HAc added = 0.1149 + 0.043756 = 0.1587 mol HAc

Mol/L of pure HAc = 1049 g HAc/L × (1 mol HAc/60.05 g) = 17.4$\underline{6}$7 mol/L

L of pure HAc needed = 0.1587 mol HAc × (L/17.467 mol) = 0.009$\underline{0}$8 L (9.1 mL)

17. SOLUBILITY AND COMPLEX-ION EQUILIBRIA

■ Answers to Review Questions

17.1   The solubility equation is $Ni(OH)_2(s) \rightleftharpoons Ni^{2+}(aq) + 2OH^-(aq)$. If the molar solubility of $Ni(OH)_2 = x$ molar, the concentrations of the ions in the solution must be x M Ni^{2+} and $2x$ M OH^-. Substituting into the equilibrium-constant expression gives

$$K_{sp} = [Ni^{2+}][OH^-]^2 = (x)(2x)^2 = 4x^3$$

17.2   Calcium sulfate is less soluble in a solution containing sodium sulfate because the increase in sulfate from the sodium sulfate causes the equilibrium composition in the equation below to shift to the left:

$$CaSO_4(s) \rightleftharpoons Ca^{2+}(aq) + SO_4^{2-}(aq)$$

The result is a decrease in both the calcium ion and the calcium sulfate concentrations.

17.3   In order to predict whether or not PbI_2 will precipitate when lead nitrate and potassium iodide are mixed, the concentrations of Pb^{2+} and I^- after mixing would have to be calculated first (if the concentrations are not known or are not given). Then the value of Q_c, the ion product, would have to be calculated for PbI_2. Finally, Q_c would have to be compared with the value of K_{sp}. If $Q_c > K_{sp}$, then a precipitate will form at equilibrium. If Q_c is $\leq$ than K_{sp}, no precipitate will form.

17.4   Barium fluoride, normally insoluble in water, dissolves in dilute hydrochloric acid because the fluoride ion, once it forms, reacts with the hydronium ion to form weakly ionized HF:

$$BaF_2(s) \rightleftharpoons Ba^{2+}(aq) + 2\,F^-(aq)\ [+ 2\,H_3O^+ \rightarrow 2\,HF + 2\,H_2O]$$

17.5   Metal ions such as Pb^{2+} and Zn^{2+} are separated by controlling the $[S^{2-}]$ in a solution of saturated H_2S by means of adjusting the pH correctly. Because the K_{sp} of 2.5×10^{-27} for PbS is smaller than the K_{sp} of 1.1×10^{-21} for ZnS, the pH can be adjusted to make the $[S^{2-}]$ just high enough to precipitate PbS without precipitating ZnS.

17.6   When a small amount of NaOH is added to a solution of $Al_2(SO_4)_3$, a precipitate of $Al(OH)_3$ forms at first. As more NaOH is added, the excess hydroxide ion reacts further with the insoluble $Al(OH)_3$, forming a soluble complex ion of $Al(OH)_4^-$.

■ Solutions to Practice Problems

Note on significant figures: If the final answer to a solution needs to be rounded off, it is given first with one nonsignificant figure, and the last significant figure is underlined. The final answer is then rounded to the correct number of significant figures. In multiple-step problems, intermediate answers are given with at least one nonsignificant figure; however, only the final answer has been rounded off.

17.13 a. $K_{sp} = [Mg^{2+}][OH^-]^2$ b. $K_{sp} = [Sr^{2+}][CO_3^{2-}]$

   c. $K_{sp} = [Ca^{2+}]^3[AsO_4^{3-}]^2$ d. $K_{sp} = [Fe^{3+}][OH^-]^3$

17.15   Calculate the molar solubility. Then assemble the usual concentration table, and substitute the equilibrium concentrations from it into the equilibrium-constant expression.

$$\frac{0.13 \text{ g}}{0.100 \text{ L}} \times \frac{1 \text{ mol}}{413 \text{ g}} = 3.\underline{1}47 \times 10^{-3} \text{ M}$$

Conc. (M)	$Cu(IO_3)_2(s)$ ⇌	Cu^{2+}	+	$2IO_3^-$
Starting		0		0
Change		$+3.147 \times 10^{-3}$		$+2 \times 3.147 \times 10^{-3}$
Equilibrium		3.147×10^{-3}		$2 \times 3.147 \times 10^{-3}$

$K_{sp} = [Cu^{2+}][IO_3^-]^2 = (3.147 \times 10^{-3})[2(3.147 \times 10^{-3})]^2 = 1.\underline{2}47 \times 10^{-7} = 1.2 \times 10^{-7}$

17.17 Calculate the pOH from the pH, and then convert pOH to [OH⁻]. Then assemble the usual concentration table, and substitute the equilibrium concentrations from it into the equilibrium-constant expression.

$pOH = 14.00 - 10.52 = 3.48$

$[OH^-] = antilog(-pOH) = antilog(-3.48) = 3.\underline{3}11 \times 10^{-4}\ M$

$[Mg^{2+}] = [OH^-] \div 2 = (3.311 \times 10^{-4}) \div 2 = 1.655 \times 10^{-4}\ M$

Conc. (M)	Mg(OH)₂(s) ⇌	Mg²⁺	+	2OH⁻
Starting		0		0
Change		+1.655 x 10⁻⁴		+3.311 x 10⁻⁴
Equilibrium		1.655 x 10⁻⁴		3.311 x 10⁻⁴

$K_{sp} = [Mg^{2+}][OH^-]^2 = (1.655 \times 10^{-4})(3.311 \times 10^{-4})^2 = 1.\underline{8}1 \times 10^{-11} = 1.8 \times 10^{-11}$

17.19 Assemble the usual concentration table. Let x equal the molar solubility of SrCO₃. When x mol SrCO₃ dissolves in one L of solution, x mol Sr²⁺ and x mol CO₃²⁻ form.

Conc. (M)	SrCO₃(s) ⇌	Sr²⁺	+	CO₃²⁻
Starting		0		0
Change		+x		+x
Equilibrium		x		x

Substitute the equilibrium concentrations into the equilibrium-constant expression, and solve for x. Then convert to grams SrCO₃ per liter.

$[Sr^{2+}][CO_3^{2-}] = K_{sp}$

$(x)(x) = x^2 = 9.3 \times 10^{-10}$

$x = \sqrt{(9.3 \times 10^{-10})} = 3.\underline{0}4 \times 10^{-5}\ M$

$\dfrac{3.04 \times 10^{-5}\ mol}{L} \times \dfrac{147.63\ g}{1\ mol\ SrCO_3} = 4.\underline{5}0 \times 10^{-3} = 0.0045\ g/L$

17.21 Let x equal the molar solubility of $PbBr_2$. Assemble the usual concentration table, and substitute from the table into the equilibrium-constant expression.

Conc. (M)	$PbBr_2(s)$	$\rightleftharpoons$	Pb^{2+}	+	$2Br^-$
Starting			0		0
Change			+x		+2x
Equilibrium			x		2x

$[Pb^{2+}][Br^-]^2 = K_{sp}$

$(x)(2x)^2 = 4x^3 = 6.3 \times 10^{-6}$

$$x = \sqrt[3]{\frac{6.3 \times 10^{-6}}{4}} = 0.0116 = 0.012\ M$$

17.23 At the start, before any $SrSO_4$ dissolves, the solution contains $0.23\ M\ SO_4^{2-}$. At equilibrium, x mol of solid $SrSO_4$ dissolves to yield x mol Sr^{2+} and x mol SO_4^{2-}. Assemble the usual concentration table, and substitute the equilibrium concentrations into the equilibrium-constant expression. As an approximation, assume x is negligible compared to $0.23\ M\ SO_4^{2-}$.

Conc. (M)	$SrSO_4(s)$	$\rightleftharpoons$	Sr^{2+}	+	SO_4^{2-}
Starting			0		0.23
Change			+x		+x
Equilibrium			x		0.23 + x

$[Sr^{2+}][SO_4^{2-}] = K_{sp}$

$(x)(0.23 + x) \cong (x)(0.23) \cong 2.5 \times 10^{-7}$

$$x \cong \frac{2.5 \times 10^{-7}}{0.23} = 1.086 \times 10^{-6}\ M$$

$$\frac{1.086 \times 10^{-6}\ mol}{L} \times \frac{183.69\ g}{1\ mol\ SrSO_4} = 1.99 \times 10^{-4} = 2.0 \times 10^{-4}\ g/L$$

Note that adding x to $0.23\ M$ will not change it (to two significant figures), so x is negligible compared to $0.23\ M$.

17.25 Calculate the value of K_{sp} from the solubility using the concentration table. Then using the common-ion calculation, assemble another concentration table. Use 0.020 M NaF as the starting concentration of F⁻ ion. Substitute the equilibrium concentrations from the table into the equilibrium-constant expression. As an approximation, assume $2x$ is negligible compared to 0.020 M F⁻ ion.

$$\frac{0.016 \text{ g}}{L} \times \frac{1 \text{ mol MgF}_2}{62.31 \text{ g}} = 2.\underline{5}67 \times 10^{-4} \, M$$

Conc. (M)	MgF$_2$(s)	⇌	Mg^{2+}	+	2F⁻
Starting			0		0
Change			+2.$\underline{5}$67 x 10⁻⁴		+2 x 2.$\underline{5}$67 x 10⁻⁴
Equilibrium			2.$\underline{5}$67 x 10⁻⁴		2 x 2.$\underline{5}$67 x 10⁻⁴

$K_{sp} = [\text{Mg}^{2+}][\text{F}^-]^2 = (2.\underline{5}67 \times 10^{-4})[2(2.\underline{5}67 \times 10^{-4})]^2 = 6.\underline{7}72 \times 10^{-11}$

Now, use K_{sp} to calculate the molar solubility of MgF$_2$.

Conc. (M)	MgF$_2$(s)	⇌	Mg^{2+}	+	2F⁻
Starting			0		0.020
Change			+x		+2x
Equilibrium			x		0.020 + 2x

$[\text{Mg}^{2+}][\text{F}^-]^2 = K_{sp}$

$(x)(0.020 + 2x)^2 \cong (x)(0.020)^2 = 6.\underline{7}72 \times 10^{-11}$

$$x \cong \frac{6.772 \times 10^{-11}}{(0.020)^2} = 1.\underline{6}93 \times 10^{-7} \, M$$

$$\frac{1.693 \times 10^{-7} \text{ mol}}{L} \times \frac{62.31 \text{ g}}{1 \text{ mol MgF}_2} = 1.\underline{0}54 \times 10^{-5} = 1.1 \times 10^{-5} \text{ g/L}$$

17.27 a. Calculate Q_c, the ion product of the solution, using the concentrations in the problem as the concentrations present after mixing and assuming no precipitation. Then compare Q_c with K_{sp} to determine whether precipitation has occurred. Start by defining the ion product with brackets as used for the definition of K_{sp}; then use parentheses for the concentrations.

$$Q_c = [Ba^{2+}][F^-]^2$$

$$Q_c = (0.020)(0.015)^2 = 4.\underline{5}0 \times 10^{-6}$$

Since $Q_c > K_{sp}$ (1.0×10^{-6}), precipitation will occur.

   b. Calculate Q_c, the ion product of the solution, using the concentrations in the problem as the concentrations present after mixing and assuming no precipitation. Then compare Q_c with K_{sp} to determine whether precipitation has occurred. Start by defining the ion product with brackets as used for the definition of K_{sp}; then use parentheses for the concentrations.

$$Q_c = [Mg^{2+}][CO_3^{2-}]$$

$$Q_c = (0.0012)(0.041) = 8.\underline{6}1 \times 10^{-5}$$

Since $Q_c > K_{sp}$ (1.0×10^{-5}), precipitation will occur.

17.29 Calculate the concentrations of Mg^{2+} and OH^-, assuming no precipitation. Use a total volume of 1.0 L + 1.0 L, or 2.0 L. (Note the concentrations are halved when the volume is doubled.)

$$[Mg^{2+}] = \frac{\dfrac{0.0020 \text{ mol}}{L} \times 1.0 \text{ L}}{2.0 \text{ L}} = 1.\underline{0}0 \times 10^{-3} \text{ M}$$

$$[OH^-] = \frac{\dfrac{0.00010 \text{ mol}}{L} \times 1.0 \text{ L}}{2.0 \text{ L}} = 5.\underline{0}0 \times 10^{-5} \text{ M}$$

Calculate the ion product, and compare it to K_{sp}.

$$Q_c = (Mg^{2+})(OH^-)^2 = (1.00 \times 10^{-3})(5.00 \times 10^{-5})^2 = 2.\underline{5}0 \times 10^{-12}$$

Because Q_c is less than the K_{sp} of 1.8×10^{-11}, no precipitation occurs, and the solution is unsaturated.

17.31 A mixture of $CaCl_2$ and K_2SO_4 can only precipitate $CaSO_4$ because KCl is soluble. Use the K_{sp} expression to calculate the $[Ca^{2+}]$ needed to just begin precipitating the $0.020\ M\ SO_4^{2-}$ (in essentially a saturated solution). Then convert to moles.

$$[Ca^{2+}][SO_4^{2-}] = K_{sp} = 2.4 \times 10^{-5}$$

$$[Ca^{2+}] = \frac{K_{sp}}{[SO_4^{2-}]} = \frac{2.4 \times 10^{-5}}{2.0 \times 10^{-2}} = 1.\underline{2}0 \times 10^{-3}\ M$$

The number of moles in 1.5 L of this calcium-containing solution is (mol $CaCl_2$ = mol Ca^{2+}):

$$1.5\ L \times (1.20 \times 10^{-3})\ mol/L = 1.\underline{8}0 \times 10^{-3} = 0.0018\ mol\ CaCl_2$$

17.33 The net ionic equation is

$$CaC_2O_4(s) + 2\ H_3O^+(aq) \rightleftharpoons Ca^{2+}(aq) + H_2C_2O_4(aq) + 2\ H_2O(l)$$

17.35 Calculate the value of K for the reaction of H_3O^+ with both the SO_4^{2-} and F^- anions as they form by the slight dissolving of the insoluble salts. These constants are the reciprocals of the K_a values of the conjugate acids of these anions.

$$F^-(aq) + H_3O^+(aq) \rightleftharpoons HF(aq) + H_2O(l)$$

$$K = \frac{1}{K_a} = \frac{1}{6.8 \times 10^{-4}} = 1.\underline{4}7 \times 10^3$$

$$SO_4^{2-}(aq) + H_3O^+(aq) \rightleftharpoons HSO_4^-(aq) + H_2O(l)$$

$$K = \frac{1}{K_{a2}} = \frac{1}{1.1 \times 10^{-2}} = 9\underline{0}.9$$

Because K for the fluoride ion is relatively larger, more CaF_2 will dissolve in acid than $CaSO_4$.

17.37 The equation is

$$Cu^+(aq) + 2\ CN^-(aq) \rightleftharpoons Cu(CN)_2^-(aq)$$

The K_f expression is

$$K_f = \frac{[Cu(CN)_2^-]}{[Cu^+][CN^-]^2} = 1.0 \times 10^{16}$$

17.39 Assume the only $[Ag^+]$ is that in equilibrium with the $Ag(CN)_2^-$ formed from Ag^+ and CN^-. (In other words, assume all the 0.015 M Ag^+ reacts with CN^- to form 0.015 M $Ag(CN)_2^-$.) Subtract the CN^- that forms the 0.015 M $Ag(CN)_2^-$ from the initial 0.100 M CN^-. Use this as the starting concentration of CN^- for the usual concentration table.

$$[0.100 \text{ } M \text{ NaCN} - (2 \times 0.015 \text{ } M)] = 0.070 \text{ } M \text{ starting } CN^-$$

Conc. (M)	$Ag(CN)_2^-$ $\rightleftharpoons$	Ag^+	+	$2CN^-$
Starting	0.015	0		0.070
Change	$-x$	$+x$		$+2x$
Equilibrium	$0.015 - x$	x		$0.070 + 2x$

Even though this reaction is the opposite of the equation for the formation constant, the formation-constant expression can be used. Simply substitute all the exact equilibrium concentrations into the formation-constant expression; then simplify the exact equation by assuming x is negligible compared to 0.015, and $2x$ is negligible compared to 0.070.

$$K_f = \frac{[Ag(CN)_2^-]}{[Ag^+][CN^-]^2} = \frac{(0.015 - x)}{(x)(0.070 + 2x)^2} \cong \frac{(0.015)}{(x)(0.070)^2} \cong 5.6 \times 10^{18}$$

Rearrange and solve for $x = [Ag^+]$:

$$x \cong (0.015) \div [(5.6 \times 10^{18})(0.070)^2] \cong 5.\underline{4}6 \times 10^{-19} = 5.5 \times 10^{-19} \text{ } M$$

17.41 The Pb^{2+}, Cd^{2+}, and Sr^{2+} ions can be separated in two steps: (1) Add HCl to precipitate only the Pb^{2+} as $PbCl_2$, leaving the others in solution. (2) After pouring the solution away from the precipitate, add 0.3 M HCl and H_2S to precipitate only the CdS away from the Sr^{2+} ion, whose sulfide is soluble under these conditions.

■ Solutions to General Problems

17.43 Assemble the usual concentration table. Let x equal the molar solubility of $PbSO_4$. When x mol $PbSO_4$ dissolves in one L of solution, x mol Pb^{2+} and x mol SO_4^{2-} form.

Conc. (M)	$PbSO_4(s)$ $\rightleftharpoons$	Pb^{2+}	+	SO_4^{2-}
Starting		0		0
Change		$+x$		$+x$
Equilibrium		x		x

(continued)

Substitute the equilibrium concentrations into the equilibrium-constant expression, and solve for x.

$$[Pb^{2+}][SO_4^{2-}] = K_{sp}$$

$$(x)(x) = x^2 = 1.7 \times 10^{-8}$$

$$x = \sqrt{(1.7 \times 10^{-8})} = 1.\underline{3}03 \times 10^{-4} = 1.3 \times 10^{-4} \; M$$

17.45 Calculate the pOH from the pH, and then convert pOH to $[OH^-]$. Then assemble the usual concentration table, and substitute the equilibrium concentrations from it into the equilibrium-constant expression.

$$pOH = 14.00 - 8.80 = 5.20$$

$$[OH^-] = \text{antilog} (-pOH) = \text{antilog} (-5.20) = 6.\underline{3}09 \times 10^{-6} \; M$$

Conc. (M)	$Mg(OH)_2(s)$	$\rightleftharpoons$	Mg^{2+}	+	$2OH^-$
Starting			0		0
Change			+x		—
Equilibrium			x		6.309×10^{-6}

$$K_{sp} = [Mg^{2+}][OH^-]^2 = (x)(6.309 \times 10^{-6})^2 = 1.8 \times 10^{-11}$$

$$x = \frac{1.8 \times 10^{-11}}{(6.309 \times 10^{-6})^2} = 0.4\underline{5}2 = 0.45 \; M$$

$$\frac{0.452 \text{ mol}}{L} \times \frac{58.3 \text{ g } Mg(OH)_2}{1 \text{ mol}} = 2\underline{6}.3 = 26 \text{ g/L}$$

17.47 Let x equal the change in M of Mg^{2+} and 0.10 M equal the starting OH^- concentration. Then assemble the usual concentration table, and substitute the equilibrium concentrations from it into the equilibrium-constant expression. Assume $2x$ is negligible compared to 0.10 M, and perform an approximate calculation.

Conc. (M)	$Mg(OH)_2(s)$	$\rightleftharpoons$	Mg^{2+}	+	$2OH^-$
Starting			0		0.10
Change			+x		+2x
Equilibrium			x		0.10 + 2x

(continued)

$$K_{sp} = [Mg^{2+}][OH^-]^2 = (x)(0.10 + 2x)^2 \cong (x)(0.10)^2 \cong 1.8 \times 10^{-11}$$

$$x = \frac{1.8 \times 10^{-11}}{(0.10)^2} = 1.\underline{8}0 \times 10^{-9} = 1.8 \times 10^{-9} \, M$$

17.49   To begin precipitation, you must add just slightly more sulfate ion than that required to give a saturated solution. Use the K_{sp} expression to calculate the $[SO_4^{2-}]$ needed to just begin precipitating the $0.0030 \, M \, Ca^{2+}$.

$$[Ca^{2+}][SO_4^{2-}] = K_{sp} = 2.4 \times 10^{-5}$$

$$[SO_4^{2-}] = \frac{K_{sp}}{[Ca^{2+}]} = \frac{2.4 \times 10^{-5}}{3.0 \times 10^{-3}} = 8.\underline{0}0 \times 10^{-3} \, M$$

When the sulfate ion concentration slightly exceeds $8.0 \times 10^{-3} \, M$, precipitation begins.

17.51   A mixture of $AgNO_3$ and $NaCl$ can precipitate only $AgCl$ because $NaNO_3$ is soluble. Use the K_{sp} expression to calculate the $[Cl^-]$ needed to prepare a saturated solution (just before precipitating the $0.0015 \, M \, Ag^+$). Then convert to moles and finally to grams.

$$[Ag^+][Cl^-] = K_{sp} = 1.8 \times 10^{-10}$$

$$[Cl^-] = \frac{K_{sp}}{[Ag^+]} = \frac{1.8 \times 10^{-10}}{1.5 \times 10^{-3}} = 1.\underline{2}0 \times 10^{-7} \, M$$

The number of moles and grams in 0.785 L (785 mL) of this chloride-containing solution is

$$0.785 \, L \times (1.20 \times 10^{-7} \, mol/L) = 9.\underline{4}2 \times 10^{-8} \, mol \, Cl^- = 9.\underline{4}2 \times 10^{-8} \, mol \, NaCl$$

$$(9.42 \times 10^{-7} \, mol \, NaCl) \times (58.5 \, g \, NaCl/1 mol) = 5.\underline{5}1 \times 10^{-6} = 5.5 \times 10^{-6} \, g \, NaCl$$

This amount of NaCl is too small to weigh on a balance.

17.53 Obtain the overall equilibrium constant for this reaction from the product of the individual equilibrium constants of the two individual equations whose sum gives this equation:

$AgBr(s)$ ⇌ $Ag^+(aq)$ + $Br^-(aq)$ $K_{sp} = 5.0 \times 10^{-13}$

$Ag^+(aq) + 2NH_3(aq)$ ⇌ $Ag(NH_3)_2^+(aq)$ $K_f = 1.7 \times 10^7$

$AgBr(s) + 2NH_3(aq)$ ⇌ $Ag(NH_3)_2^{2+}(aq)$ + $Br^-(aq)$

$$K_c = K_{sp} \times K_f = 8.\underline{5}0 \times 10^{-6}$$

Assemble the usual table using 5.0 M as the starting concentration of NH_3 and x as the unknown concentration of $Ag(NH_3)_2^+$ formed.

Conc. (M)	AgBr(s)	+	2NH₃	⇌	Ag(NH₃)₂⁺	+	Br⁻
Starting			5.0		0		0
Change			-2x		+x		+x
Equilibrium			5.0 - 2x		x		x

The equilibrium-constant expression can now be used. Simply substitute all the exact equilibrium concentrations into the equilibrium-constant expression; it will not be necessary to simplify the equation because taking the square root of both sides removes the x^2 term.

$$K_c = \frac{[Ag(NH_3)_2^+][Br^-]}{[NH_3]^2} = \frac{x^2}{(5.0 - 2x)^2} = 8.\underline{5}0 \times 10^{-6}$$

Take the square root of both sides of the two right-hand terms, and solve for x.

$$\frac{x}{(5.0 - 2x)} = 2.\underline{9}15 \times 10^{-3}$$

$$x + (5.8 \times 10^{-3})x = 1.\underline{4}575 \times 10^{-2}$$

$$x = 1.\underline{4}49 \times 10^{-2} = 1.4 \times 10^{-2}\ M \text{ (the molar solubility of AgBr in 5.0 } M\ NH_3)$$

17.55 The OH^- formed by ionization of NH_3 (to NH_4^+ and OH^-) is a common ion that will precipitate Mg^{2+} as the slightly soluble $Mg(OH)_2$ salt. The simplest way to treat the problem is to calculate the $[OH^-]$ of 0.10 $M\ NH_3$ before the soluble Mg^{2+} salt is added. As the soluble Mg^{2+} salt is added, the $[Mg^{2+}]$ will increase until the solution is saturated with respect to $Mg(OH)_2$ (the next Mg^{2+} ions to be added will precipitate). Calculate the $[Mg^{2+}]$ at the point at which precipitation begins.

(continued)

To calculate the [OH⁻] of 0.10 M NH₃, let x equal the mol/L of NH₃ that ionize, forming x mol/L of NH₄⁺ and x mol/L of OH⁻ and leaving (0.10 - x) M NH₃ in solution. We can summarize the situation in tabular form:

Conc. (M)	NH₃	+	H₂O	⇌	NH₄⁺	+	OH⁻
Starting	0.10				~0		0
Change	-x				+x		+x
Equilibrium	0.10 - x				x		x

The equilibrium-constant equation is:

$$K_c = \frac{[NH_4^+][OH^-]}{[NH_3]} = \frac{x^2}{(0.10 - x)} \cong \frac{x^2}{(0.10)}$$

The value of x can be obtained by rearranging and taking the square root:

$$[OH^-] \cong \sqrt{1.8 \times 10^{-5} \times 0.10} = 1.\underline{3}4 \times 10^{-3} \, M$$

Note that (0.10 - x) is not significantly different from 0.10, so x can be ignored in the (0.10 - x) term.

Now use the K_{sp} of Mg(OH)₂ to calculate the [Mg²⁺] of a saturated solution of Mg(OH)₂, which essentially will be the Mg²⁺ ion concentration when Mg(OH)₂ begins to precipitate.

$$[Mg^{2+}] = \frac{K_{sp}}{[OH^-]^2} = \frac{1.8 \times 10^{-11}}{(1.34 \times 10^{-3})^2} = 1.\underline{0}0 \times 10^{-5} = 1.0 \times 10^{-5} \, M$$

17.57 a. Use the solubility information to calculate K_{sp}. The reaction is

$$Cu(IO_3)_2(s) \rightleftharpoons Cu^{2+}(aq) + 2IO_3^-(aq)$$
$$2.7 \times 10^{-3} \qquad 2 \times (2.7 \times 10^{-3})$$

$$K_{sp} = [Cu^{2+}][IO_3^-]^2 = [2.7 \times 10^{-3}][2 \times (2.7 \times 10^{-3})]^2 = 7.\underline{8}7 \times 10^{-8}$$

Set up an equilibrium. The reaction is

$$Cu(IO_3)_2(s) \rightleftharpoons Cu^{2+}(aq) + 2IO_3^-(aq)$$
$$y \qquad 0.35 + 2y \approx 0.35$$

(continued)

$K_{sp} = [y][0.35]^2 = 7.87 \times 10^{-8}$

Molar solubility $= y = 6.\underline{4}2 \times 10^{-7} = 6.4 \times 10^{-7}\ M$

b. Set up an equilibrium. The reaction is

$$Cu(IO_3)_2(s) \rightleftharpoons Cu^{2+}(aq) \quad + \quad 2IO_3^-(aq)$$

$$y + 0.35 \approx 0.35 \qquad 2y$$

$K_{sp} = [0.35][2y]^2 = 7.87 \times 10^{-8}$

Molar solubility $= y = 2.\underline{3}7 \times 10^{-4} = 2.4 \times 10^{-4}\ M$

c. Yes, $Cu(IO_3)_2$ is a 1:2 electrolyte. It takes two IO_3^- ions to combine with one Cu^{2+} ion. The IO_3^- ion is involved as a square term in the K_{sp} expression.

17.59 a. Set up an equilibrium. The reaction and equilibrium-constant expression are

$$PbI_2(s) \rightleftharpoons Pb^{2+}(aq) + 2I^-(aq) \qquad K_{sp} = [Pb^{2+}][I^-]^2 = 6.5 \times 10^{-9}$$

The Pb^{2+} ion concentration is 0.0150 M. Plug this in, and solve for the iodine ion concentration.

$$[I^-] = \sqrt{\frac{6.5 \times 10^{-9}}{0.0150}} = 6.\underline{5}8 \times 10^{-4} = 6.6 \times 10^{-4}\ M$$

b. Solve the equilibrium-constant expression for the lead ion concentration.

$$[Pb^{2+}] = \frac{6.5 \times 10^{-9}}{(2.0 \times 10^{-3})^2} = 1.\underline{6}3 \times 10^{-3}\ M$$

The percent of the lead(II) ion remaining in solution is

$$\text{Percent } Pb^{2+} \text{ remaining} = \frac{1.63 \times 10^{-3}}{0.0150} \times 100\% = 1\underline{0}.9 = 11 \text{ percent}$$

17.61 a. The reaction and equilibrium-constant expression are

$$Co(OH)_2(s) \rightleftharpoons Co^{2+}(aq) + 2OH^-(aq) \qquad K_{sp} = [Co^{2+}][OH^-]^2$$

From the molar solubility, $[Co^{2+}] = 5.4 \times 10^{-6}\ M$ and $[OH^-] = 2 \times (5.4 \times 10^{-6})\ M$. Therefore,

$$K_{sp} = [5.4 \times 10^{-6}][2 \times (5.4 \times 10^{-6})]^2 = 6.\underline{3}0 \times 10^{-16} = 6.3 \times 10^{-16}$$

(continued)

b. From the pOH (14 - pH), the $[OH^-] = 10^{-3.57} = 2.\underline{69} \times 10^{-4}$ M. The molar solubility is equal to the cobalt ion concentration at equilibrium.

$$[Co^{2+}] = \frac{K_{sp}}{[OH^-]^2} = \frac{6.3 \times 10^{-16}}{(2.69 \times 10^{-4})^2} = 8.\underline{71} \times 10^{-9} = 8.7 \times 10^{-9} \ M$$

c. The common ion effect (OH^-) in part (b) decreases the solubility of Co^{2+}.

17.63 a. The moles of NH_4Cl (molar mass 53.49 g/mol) are given by

$$mol \ NH_4Cl = 26.7 \ g \times \frac{1 \ mol \ NH_4Cl}{53.49 \ g \ NH_4Cl} = 0.49\underline{9}2 \ mol$$

The reaction and equilibrium-constant expression are

$$NH_3 + H_2O \rightleftharpoons NH_4^+ + OH^- \qquad K_b = \frac{[NH_4^+][OH^-]}{[NH_3]} = 1.8 \times 10^{-5}$$

Solve the equilibrium-constant expression for $[OH^-]$. The molarity of the NH_4Cl is 0.49$\underline{9}$2 mol/1.0 L = 0.49$\underline{9}$2 M.

$$[OH^-] = \frac{(1.8 \times 10^{-5})(4.2)}{(0.4992)} = 1.\underline{51} \times 10^{-4} = 1.5 \times 10^{-4} \ M$$

b. The reaction and equilibrium-constant expression are

$$Mg(OH)_2 \rightleftharpoons Mg^{2+} + 2OH^- \qquad K_{sp} = [Mg^{2+}][OH^-]^2 = 1.8 \times 10^{-11}$$

Solving for $[Mg^{2+}]$ gives

$$[Mg^{2+}] = \frac{(1.8 \times 10^{-11})}{(1.51 \times 10^{-4})^2} = 7.\underline{89} \times 10^{-4} = 7.9 \times 10^{-4} \ M$$

The percent Mg^{2+} that has been removed is given by

$$Percent \ Mg^{2+} \ removed = \frac{0.075 - 7.89 \times 10^{-4}}{0.075} \times 100\% = 98.\underline{9} = 99 \ percent$$

■ Solutions to Cumulative-Skills Problems

17.65 Begin by solving for $[H_3O^+]$ in the buffer. Ignoring changes in $[HCHO_2]$ as a result of ionization in the buffer, you obtain

$$[H_3O^+] \cong 1.7 \times 10^{-4} \times \frac{0.45 \text{ M}}{0.20 \text{ M}} = 3.\underline{8}25 \times 10^{-4} \text{ } M$$

You should verify that this approximation is valid (you obtain the same result from the Henderson-Hasselbalch equation). The equilibrium for the dissolution of CaF_2 in acidic solution is obtained by subtracting twice the acid ionization of HF from the solubility equilibrium of CaF_2:

$$CaF_2(s) \rightleftharpoons Ca^{2+}(aq) + 2F^-(aq) \qquad K_{sp}$$

$$2H_3O^+(aq) + 2F^-(aq) \rightleftharpoons 2HF(aq) + 2H_2O(l) \qquad 1/(K_a)^2$$

$$2H_3O^+(aq) + CaF_2(s) \rightleftharpoons Ca^{2+}(aq) + 2HF(aq) + 2H_2O(l) \quad K_c = K_{sp}/(K_a)^2$$

Therefore, $K_c = (3.4 \times 10^{-11}) \div (6.8 \times 10^{-4})^2 = 7.\underline{3}5 \times 10^{-5}$. In order to solve the equilibrium-constant equation, you require the concentration of HF, which you obtain from the acid-ionization constant for HF.

$$K_a = \frac{[H_3O^+][F^-]}{[HF]}$$

$$6.8 \times 10^{-4} = 3.825 \times 10^{-4} \times \frac{[F^-]}{[HF]}$$

$$\frac{[F^-]}{[HF]} = 1.778, \text{ or } [F^-] = 1.\underline{7}78 \text{ [HF]}$$

Let x be the solubility of CaF_2 in the buffer. Then $[Ca^{2+}] = x$, and $[F^-] + [HF] = 2x$. Substituting from the previous equation, you obtain

$$2x = 1.778[HF] + [HF] = 2.778[HF], \text{ or } [HF] = 2x/2.\underline{7}78$$

You can now substitute for $[H_3O^+]$ and $[HF]$ into the equation for K_c.

$$K_c = \frac{[Ca^{2+}][HF]^2}{[H_3O^+]^2} = \frac{x(2x/2.778)^2}{(3.825 \times 10^{-4})^2} = 7.35 \times 10^{-5}$$

(continued)

$$7.35 \times 10^{-5} = (3.543 \times 10^6)x^3$$

$$x^3 = 2.075 \times 10^{-11}$$

$$x = 2.\underline{7}48 \times 10^{-4} = 2.7 \times 10^{-4} \ M$$

17.67 The net ionic equation is

$$Ba^{2+}(aq) + 2OH^-(aq) + Mg^{2+}(aq) + SO_4^{2-}(aq) \rightleftharpoons BaSO_4(s) + Mg(OH)_2(s)$$

Start by calculating the mol/L of each ion after mixing and before precipitation. Use a total volume of 0.0450 + 0.0670 L = 0.112 L.

M of SO_4^{2-} and Mg^{2+} = (0.350 mol/L x 0.0670 L) ÷ 0.112 L = 0.20$\underline{9}$4 M

M of Ba^{2+} = (0.250 mol/L x 0.0450 L) ÷ 0.112 L = 0.10$\underline{0}$4 M

M of OH^- = (2 x 0.250 mol/L x 0.0450 L) ÷ 0.112 L = 0.20$\underline{0}$89 M

Assemble a table showing the precipitation of $BaSO_4$ and $Mg(OH)_2$.

Conc. (M)	Ba^{2+}	+	SO_4^{2-}	+	Mg^{2+}	+	$2OH^-$
						$\rightleftharpoons$	$BaSO_4(s)$ + $Mg(OH)_2(s)$
Starting	0.1004		0.2094		0.2094		0.20089
Change	-0.1004		-0.1004		-0.1004		-0.20089
Equilibrium	0.0000		0.1090		0.1090		0.0000

To calculate $[Ba^{2+}]$, use the K_{sp} expression for $BaSO_4$.

$$[Ba^{2+}] = \frac{1.1 \times 10^{-10}}{0.1090 \ M \ SO_4^{2-}} = 1.\underline{0}09 \times 10^{-9} = 1.0 \times 10^{-9} \ M$$

$$[SO_4^{2-}] = 0.10\underline{9}0 = 0.109 \ M$$

Calculate the $[OH^-]$ using the K_{sp} expression for $Mg(OH)_2$:

$$[OH^-] = \sqrt{\frac{1.8 \times 10^{-11}}{0.1090 \ M \ Mg^{2+}}} = 1.\underline{2}8 \times 10^{-5} = 1.3 \times 10^{-5} \ M$$

$$[Mg^{2+}] = 0.10\underline{9}0 = 0.109 \ M$$

18. THERMODYNAMICS AND EQUILIBRIUM

■ Answers to Review Questions

18.1   A spontaneous process is a chemical and/or a physical change that occurs by itself without the continuing intervention of an outside agency. Three examples are (1) a rock on a hilltop rolls down, (2) heat flows from a hot object to a cold one, and (3) iron rusts in moist air. Three examples of nonspontaneous processes are (1) a rock rolls uphill by itself, (2) heat flows from a cold object to a hot one, and (3) rust is converted to iron and oxygen.

18.2   Because the energy is more dispersed in liquids than in solids, liquid benzene contains more entropy than does the same quantity of frozen benzene.

18.3   The second law of thermodynamics states that, for a spontaneous process, the total entropy of a system and its surroundings always increases. As stated in Section 19.2, a spontaneous process actually creates energy dispersal, or entropy.

18.4   The standard entropy of hydrogen gas at 25°C can be obtained by starting near 0.0 K as a reference point because the entropy of perfect crystals of hydrogen there is almost zero. Then warm to room temperature in small increments, and calculate $\Delta S°$ for each incremental temperature change (say, 2 K) by dividing the heat absorbed by the average temperature (1 K is used as the average for 0 K to 2 K), and also take into account the entropy increases that accompany a phase change.

18.5   To predict the sign of $\Delta S°$, look for a change, Δn_{gas}, in the number of moles of gas. If there is an increase in moles of gas in the products (Δn_{gas} is positive), then $\Delta S°$ should be positive. A decrease in moles of gas in the products suggests $\Delta S°$ should be negative.

18.6   Free energy, G, equals $H - TS$; that is, it is the difference between the enthalpy of a system and the product of temperature and entropy. The free-energy change, ΔG, equals $\Delta H - T\Delta S$.

18.7   The standard free-energy change, $\Delta G°$, equals $\Delta H° - T\Delta S°$; that is, it is the difference between the standard enthalpy change of a system and the product of temperature and the standard entropy change of a system. The standard free-energy change of formation is the free-energy change when one mole of a substance is formed from its elements in their stable states at one atm and at a standard temperature, usually 25°C.

18.8   If $\Delta G°$ for a reaction is negative, the equation for the reaction is spontaneous in the direction written; that is, the reactants form the products as written. If it is positive, then the equation as written is nonspontaneous.

18.9   When gasoline burns in an automobile engine, the change in free energy shows up as useful work. Gasoline, a mixture of hydrocarbons such as C_8H_{18} or octane, burns to yield energy, gaseous CO_2, and gaseous H_2O.

18.10   As a spontaneous reaction proceeds, the free energy decreases until equilibrium is reached at a minimum ΔG. See the diagram below.

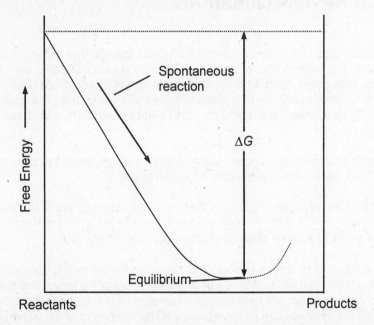

18.11   The four combinations are as follows: (1) A negative $\Delta H°$ and a positive $\Delta S°$ always give a negative $\Delta G°$ and a spontaneous reaction. (2) A positive $\Delta H°$ and a negative $\Delta S°$ always give a positive $\Delta G°$ and a nonspontaneous reaction. (3) A negative $\Delta H°$ and a negative $\Delta S°$ may give a negative or a positive $\Delta G°$. At low temperatures, $\Delta G°$ will usually be negative and the reaction spontaneous; at high temperatures, $\Delta G°$ will usually be positive and the reaction nonspontaneous. (4) A positive $\Delta H°$ and a positive $\Delta S°$ may give a negative or a positive $\Delta G°$. At low temperatures, $\Delta G°$ will usually be positive and the reaction nonspontaneous; at high temperatures, $\Delta G°$ will usually be negative and the reaction spontaneous.

18.12 You can estimate the temperature at which a nonspontaneous reaction becomes spontaneous by substituting zero for $\Delta G°$ into the equation $\Delta G° = \Delta H° - T\Delta S°$ and then solving for T using the form $T = \Delta H°/\Delta S°$.

■ Solutions to Practice Problems

Note on significant figures: The mol unit is omitted from all thermodynamic parameters such as $S°$, $\Delta S°$, etc. If the final answer to a solution needs to be rounded off, it is given first with one nonsignificant figure, and the last significant figure is underlined. The final answer is then rounded to the correct number of significant figures. In multiple-step problems, intermediate answers are given with at least one nonsignificant figure; however, only the final answer has been rounded off.

18.17 The values of q and w are -82 J and 29 J, respectively.

$$\Delta U = q + w = (-82 \text{ J}) + 29 \text{ J} = -53 \text{ J}$$

18.19 First determine the enthalpy change for the reaction of 1.20 mol of $CHCl_3$.

$$\Delta H = 1.20 \text{ mol} \times \frac{29.6 \text{ kJ}}{1 \text{ mol}} = 35.5\underline{2} \text{ kJ} = 3.5\underline{5}2 \times 10^4 \text{ J}$$

Use the equilibrium relation between ΔS and ΔH_{vap} at the boiling point (61.2°C = 334.4 K):

$$\Delta S = \frac{\Delta H}{T} = \frac{3.552 \times 10^4 \text{J}}{334.4 \text{ K}} = 10\underline{6}.2 = 106 \text{ J/K}$$

18.21 First determine the enthalpy change for the condensation of 1.00 mol of $CH_3OH(l)$. $\Delta H_{cond} = -\Delta H_{vap} = -38.0 \text{ kJ/mol}$.

The entropy change for this condensation at 25°C (298 K) is

$$\Delta S = \frac{\Delta H_{cond}}{T} = \frac{-3.80 \times 10^4 \text{J}}{298 \text{ K}} = -12\underline{7}.51 \text{ J/K}$$

The entropy of one mole of liquid is calculated using the entropy of one mole of vapor, 255 J/(mol•K).

$$S_{liq} = S_{vap} + \Delta S_{cond} = 255 \text{ J/K} + (-127.51 \text{ J/K}) = 12\underline{7}.48 = 127 \text{ J/K}$$

18.23 a. $\Delta S°$ is negative because there is a decrease in moles of gas (Δn_{gas} = -2) from three moles of gaseous reactants forming one mole of gaseous product. (Entropy decreases.)

b. $\Delta S°$ is not predictable from the rules given. The molecules N_2, O_2 and NO are of similar size and present in equal number. There is no change in moles of gas (Δn_{gas} = 0), since two moles of gaseous reactants form two moles of gaseous products. Also there is no phase change occurring.

c. $\Delta S°$ is positive because there is an increase in moles of gas (Δn_{gas} = +1) from five moles of gaseous reactants forming six moles of gaseous products. (Entropy increases.)

18.25 The reaction and standard entropies are given below. Multiply the $S°$ values by their stoichiometric coefficients, and subtract the entropy of the reactant from the sum of the product entropies.

a. $\qquad$ 2Na(s) + $\qquad$ Cl$_2$(g) $\rightarrow$ $\qquad$ 2NaCl(s)

$S°$: $\quad$ 2 x 51.46 $\qquad$ 223.0 $\qquad$ 2 x 72.12 J/K

$\Delta S°$ = $\Sigma n S°$(products) - $\Sigma m S°$(reactants) =

$\qquad$ [(2 x 72.12) - (2 x 51.46 + 223.0)] J/K = -181.$\underline{6}$8 = -181.7 J/K

b. $\qquad$ Ag(s) + $\qquad$ 1/2 Cl$_2$(g) $\rightarrow$ $\qquad$ AgCl(s)

$S°$: $\quad$ 42.55 $\qquad$ 1/2 x 223.0 $\qquad\qquad$ 96.2 $\quad$ J/K

$\Delta S°$ = $\Sigma n S°$(products) - $\Sigma m S°$(reactants) =

$\qquad$ [(96.2) - (42.55 + 1/2 x 223.0)] J/K = -57.$\underline{8}$5 = -57.9 J/K

c. $\qquad$ CS$_2$(l) + $\qquad$ 3O$_2$(g) $\rightarrow$ CO$_2$(g) + $\qquad$ 2SO$_2$(g)

$S°$: $\quad$ 151.3 $\qquad$ 3 x 205.0 $\qquad$ 213.7 $\qquad$ 2 x 248.1 J/K

$\Delta S°$ = $\Sigma n S°$(products) - $\Sigma m S°$(reactants) =

$\qquad$ [(213.7 + 2 x 248.1) - (151.3 + 3 x 205.0)] J/K = -56.4 J/K

18.27 a. Ca(s) + 1/2O$_2$(g) $\rightarrow$ CaO(s)

b. 3/2H$_2$(g) + C(graphite) + 1/2Cl$_2$(g) $\rightarrow$ CH$_3$Cl(l)

c. 1/8S$_8$(rhombic) + H$_2$(g) $\rightarrow$ H$_2$S(g)

d. P(red) + 3/2Cl$_2$(g) $\rightarrow$ PCl$_3$(g)

18.29 Write the values of ΔG_f° multiplied by their stoichiometric coefficients below each formula; then subtract ΔG_f° of the reactant from that of the products.

a.
$$CH_4(g) \quad + \quad 2O_2(g) \quad \rightarrow \quad CO_2(g) \quad + \quad 2H_2O(g)$$

ΔG_f°:   -50.80   0   -394.4   2 x (-228.6) kJ

$\Delta G^\circ = \Sigma n\Delta G_f^\circ(\text{products}) - \Sigma m\Delta G_f^\circ(\text{reactants}) =$

[(-394.4) + 2(-228.6) - (-50.80)] kJ = -800.8 kJ

b.
$$CaCO_3(s) \quad + \quad 2H^+(aq) \quad \rightarrow \quad Ca^{2+}(aq) \quad + \quad H_2O(l) \quad + \quad CO_2(g)$$

ΔG_f°:   -1128.8   0   -553.5   -237.1   -394.4 kJ

$\Delta G^\circ = \Sigma n\Delta G_f^\circ(\text{products}) - \Sigma m\Delta G_f^\circ(\text{reactants}) =$

[(-553.5) + (-237.1) + (-394.4) - (-1128.8)] kJ = -56.2 kJ

18.31 a. Spontaneous reaction

b. Nonspontaneous reaction

c. Equilibrium mixture; significant amounts of both

18.33 Calculate ΔG° per one mol Zn(s) using the given ΔG_f° values.

$$Zn(s) \quad + \quad Cu^{2+}(aq) \quad \rightarrow \quad Zn^{2+}(aq) \quad + \quad Cu(s)$$

ΔG_f°:   0   65.52   -147.0   0 kJ

$\Delta G^\circ = [(-147.0) - (65.52)]$ kJ = -212.52 kJ/mol Zn

-212.52 kJ/mol Zn x (4.85 g ÷ 65.39 g/mol Zn) = -15.76 = -15.8 kJ

Maximum work equals ΔG° equals -15.8 kJ. Because maximum work is stipulated, no entropy is produced.

18.35 a. $K = K_p = \dfrac{P_{CO_2}P_{H_2}}{P_{CO}P_{H_2O}}$

b. $K = [Li^+]^2[OH^-]^2 \, P_{H_2}$

18.37 First calculate ΔG° using the ΔG_f° values from Appendix B.

	CO(g)	+	3H$_2$(g)	$\rightleftharpoons$	CH$_4$(g)	+	H$_2$O(g)
ΔG_f°:	-137.2		0		-50.80		-228.6 kJ

Subtract ΔG_f° of the reactants from that of the products:

ΔG° = [(-50.80) + (-228.6) - (-137.2)] kJ = -142.20 kJ

Use the rearranged form of the equation, ΔG° = - RT ln K, to solve for ln K. To get compatible units, express ΔG° in joules, and set R equal to 8.31 J/(mol•K). Substituting the numerical values into the expression gives

$$\ln K = \frac{\Delta G^\circ}{-RT} = \frac{-142.20 \times 10^3}{-8.31 \times 298} = 57.\underline{4}22$$

$$K = K_p = e^{57.422} = \underline{8}.67 \times 10^{24} = 9 \times 10^{24}$$

18.39 First calculate ΔG° using the ΔG_f° values from Appendix B.

	Fe(s)	+	Cu^{2+}(aq)	$\rightleftharpoons$	Fe^{2+}(aq)	+	Cu(s)
ΔG_f°:	0		65.52		-78.87		0 kJ

Hence

ΔG° = [(-78.87) - 65.52] kJ = -144.38 kJ

Now substitute numerical values into the equation relating ln K and ΔG°.

$$\ln K = \frac{\Delta G^\circ}{-RT} = \frac{-144.38 \times 10^3}{-8.31 \times 298} = 58.\underline{3}06$$

Therefore,

$$K = K_c = e^{58.306} = \underline{2}.10 \times 10^{25} = 2 \times 10^{25}$$

18.41 From Appendix B, you have

	C(graphite)	+	CO$_2$(g)	$\rightleftharpoons$	2CO(g)
ΔH_f°:	0		-393.5		2 x (-110.5) kJ
S°:	5.740		213.7		2 x (197.5) J/K

(continued)

Calculate $\Delta H°$ and $\Delta S°$ from these values.

$$\Delta H° = [2(-110.5) - (-393.5)] \text{ kJ} = 172.\underline{5} \text{ kJ}$$

$$\Delta S° = [2(197.5) - (5.740 + 213.7)] \text{ J/K} = 175.\underline{56} \text{ J/K}$$

Substitute $\Delta H°$, $\Delta S°$ (= 0.17556 kJ/K), and T (= 1273 K) into the equation for $\Delta G_T°$.

$$\Delta G_T° = \Delta H° - T\Delta S° = 172.5 \text{ kJ} - (1273 \text{ K})(0.17556 \text{ kJ/K}) = -50.\underline{9}87 \text{ kJ}$$

Substitute the value of $\Delta G°$ (= -50.987 x 10^3 J) into the equation relating ln K and $\Delta G°$.

$$\ln K = \frac{\Delta G°}{-RT} = \frac{-50.987 \times 10^3}{-8.31 \times 1273} = 4.8\underline{1}98$$

$$K = K_p = e^{4.8198} = 1\underline{2}3.9 = 1.2 \times 10^2$$

Because K_p is greater than one, the data predict combustion of carbon should form significant amounts of CO product at equilibrium.

18.43   First calculate $\Delta H°$ and $\Delta S°$ using the given $\Delta H_f°$ and $S°$ values.

	2NaHCO$_3$(s)	→	Na$_2$CO$_3$(s)	+	H$_2$O(g)	+	CO$_2$(g)
$\Delta H_f°$:	2 x (-950.8)		-1130.8		-241.8		-393.5 kJ
$\Delta S°$:	2 x 101.7		138.8		188.7		213.7 J/K

$$\Delta H° = [(-1130.8) + (-241.8) + (-393.5) - 2(-950.8)] \text{ kJ} = 135.5 \text{ kJ}$$

$$\Delta S° = [(138.8 + 188.7 + 213.7) - 2(101.7)] \text{ J/K} = 337.8 \text{ J/K (0.3378 kJ/K)}$$

Substitute these values into $\Delta G° = \Delta H° - T\Delta S°$; let $\Delta G° = 0$, and rearrange to solve for T.

$$T = \frac{\Delta H°}{\Delta S°} = \frac{135.5 \text{ kJ}}{0.3378 \text{ kJ/K}} = 40\underline{1}.1 = 401 \text{ K}$$

■ Solutions to General Problems

18.45   The sign of $\Delta S°$ should be positive because there is an increase in moles of gas (Δn_{gas} = +5) as the solid reactant forms five moles of gas. The reaction is endothermic, denoting a positive $\Delta H°$. The fact that the reaction is spontaneous implies the product, $T\Delta S°$, is larger than $\Delta H°$, so $\Delta G°$ is negative, as required for a spontaneous reaction.

18.47   When the liquid freezes, it releases heat: ΔH_{fus} = -69.0 J/g at 16.6°C (289.6 K). The entropy change is

$$\Delta S = \frac{\Delta H_{fus}}{T} = \frac{-69.0 \text{ J/g}}{289.8 \text{ K}} \times \frac{60.05 \text{ g}}{1 \text{ mol}} = -14.\underline{3}0 = -14.3 \text{ J/(K•mol)}$$

18.49 a. $\Delta S°$ is negative because there is a decrease in moles of gas (Δn_{gas} = -1) from one mole of gaseous reactant forming aqueous and liquid products. (Entropy decreases.)

   b. $\Delta S°$ is positive because there is an increase in moles of gas (Δn_{gas} = +5) from a solid reactant forming five moles of gas. (Entropy increases.)

   c. $\Delta S°$ is positive because there is an increase in moles of gas (Δn_{gas} = +3) from two moles of gaseous reactant forming five moles of gaseous products. (Entropy increases.)

18.51   Calculate $\Delta S°$ from the individual $S°$ values:

	$C_2H_5OH(l)$	+	$O_2(g)$	→	$CH_3COOH(l)$	+	$H_2O(l)$
$S°$:	160.7		205.0		159.8		69.95 J/K

$\Delta S° = \Sigma n S°\text{(products)} - \Sigma m S°\text{(reactants)} =$

[(159.8 + 69.95) - (160.7 + 205.0)] J/K = -135.$\underline{9}$5 = -136.0 J/K

18.53   Calculate $\Delta G°$ using the $\Delta G_f°$ values from Appendix B.

	$N_2H_4(l)$	+	$O_2(g)$	→	$2H_2O(l)$	+	$N_2(g)$
$\Delta G_f°$:	149.4		0		-237.1		0 J/K

$\Delta G° = \Sigma n \Delta G_f°\text{(products)} - \Sigma m \Delta G_f°\text{(reactants)} =$

[(2 x -237.1) - (149.4)] kJ = -623.6 kJ

Because $\Delta G°$ is negative, the reaction is spontaneous as written at 25°C.

18.55   At low (room) temperature, $\Delta G°$ or ($\Delta H° - T\Delta S°$) must be positive, but at higher temperatures, $\Delta G°$ or ($\Delta H° - T\Delta S°$) must be negative. Thus, at the higher temperatures, the $-T\Delta S°$ term must become more negative than $\Delta H°$. Thus $\Delta S°$ must be positive and so must $\Delta H°$. If either were negative, $\Delta G°$ would not become negative at higher temperatures.

18.57 First calculate $\Delta G°$ using the values in Appendix B.

$$CaF_2(s) \rightleftharpoons Ca^{2+}(aq) + 2F^-(aq)$$

$\Delta G_f°:$ -1173.5   -553.5   2 x (-262.0) kJ

Hence, $\Delta G°$ for the reaction is

$$\Delta G_f° = [-553.5 + 2(-262.0) - (-1173.5)] \text{ kJ} = 96.0 \text{ kJ}$$

Now substitute numerical values into the equation relating ln K and $\Delta G°$.

$$\ln K = \frac{\Delta G°}{-RT} = \frac{96.0 \times 10^3}{-8.31 \times 298} = -38.\underline{7}66$$

$$K = K_{sp} = e^{-38.766} = \underline{1}.45 \times 10^{-17} = 1 \times 10^{-17}$$

18.59 From Appendix B, you have

$$COCl_2(g) \rightarrow CO(g) + Cl_2(g)$$

$\Delta H_f°:$ -220.1   -110.5   0 kJ

$S°:$ 283.9   197.5   223.0 J/K

Calculate $\Delta H°$ and $\Delta S°$ from these values.

$$\Delta H° = [(-110.5) - (-220.1)] \text{ kJ} = 109.6 \text{ kJ}$$

$$\Delta S° = [197.5 + 223.0 - 283.9] \text{ J/K} = 136.6 \text{ J/K} \ (0.1366 \text{ kJ/K})$$

At 25°C: $\Delta G° = \Delta H° - T\Delta S° = 109.6 \text{ kJ} - (298 \text{ K})(0.1366 \text{ kJ/K})$
$$= 68.\underline{8}9 = 68.9 \text{ kJ}$$

At 800°C: $\Delta G_T° = \Delta H° - T\Delta S° = 109.6 \text{ kJ} - (1073 \text{ K})(0.1366 \text{ kJ/K})$
$$= -37.\underline{0}7 = -37.1 \text{ kJ}$$

Thus, $\Delta G°$ changes from a positive value and a nonspontaneous reaction at 25°C to a negative value and a spontaneous reaction at 800°C.

18.61 $\Delta H° = [-393.5 + 2(-285.8) - (-238.7)]$ kJ $= -726.4$ kJ

$\Delta G° = \Delta H° - T\Delta S°$

-702.2 kJ $= -726.4$ kJ $- (298$ K$)(\Delta S°)$

$\Delta S° = -\dfrac{(-702.2 \text{ kJ}) - (-726.4 \text{ kJ})}{298 \text{ K}} = -0.08120$ kJ/K $= -81.\underline{20}$ J/K

$\Delta S° = -81.20$ J/K $= [2(70.0) + 213.7 - (126.8) - (3/2 \text{ mol}) \times S°(O_2)]$ J/K

$S°(O_2) = 205.\underline{40} = 205.4$ J/mol•K

18.63 a. The first reaction is

	$SnO_2(s)$	+	$2H_2(g)$	→	$Sn(s)$	+	$2H_2O(g)$
$\Delta H_f°$	-580.7		0		0		-241.8 kJ
$S°$	52.3		130.6		51.55		188.7 J/K

$\Delta H° = [2(-241.8) - (-580.7)]$ kJ $= 97.\underline{1}$ kJ $= 97.\underline{1} \times 10^3$ J

$\Delta S° = [2(188.7) + 51.55 - 2(130.6) - 52.3]$ J/K $= 115.\underline{45}$ J/K

The second reaction is

	$SnO_2(s)$	+	$C(s)$	→	$Sn(s)$	+	$CO_2(g)$
$\Delta H_f°$	-580.7		0		0		-393.5 kJ
$S°$	52.3		5.740		51.55		213.7 J/K

$\Delta H° = [-393.5 - (-580.7)]$ kJ $= 187.\underline{2}$ kJ $= 18\underline{7}.2 \times 10^3$ J

$\Delta S° = [213.7 + 51.55 - 5.740 - 52.3]$ J/K $= 207.\underline{21}$ J/K

b. For H_2, at what temperature does $\Delta G = 0$?

$0 = \Delta H° - T\Delta S°$

$T = \dfrac{\Delta H}{\Delta S} = \dfrac{97.1 \times 10^3 \text{ J}}{115.45 \text{ J/K}} = 84\underline{1}.0 = 841$ K

At temperatures greater than 841 K, the reaction will be spontaneous.

(continued)

For C, at what temperature does $\Delta G = 0$?

$$T = \frac{\Delta H}{\Delta S} = \frac{187.2 \times 10^3 \text{ J}}{207.21 \text{ J/K}} = 903.4 = 903 \text{ K}$$

At temperatures greater than 903 K, the reaction will be spontaneous.

c. From a consideration of temperature, the process with hydrogen would be preferred. But hydrogen is very expensive and carbon is cheap. On this basis, carbon would be preferred. Tin is produced commercially using carbon as the reducing agent.

18.65 a. If $\Delta G°$ is negative, then K must be greater than one. Consequently, the products will predominate.

b. The molecules must have enough energy to react when they collide with each other. So it depends upon the activation energy for the reaction. Usually it is necessary to heat solids for a reaction to occur as it is difficult to have effective collisions.

18.67 a. $\Delta H° = [-1285 - (-1288.3)] \text{ kJ} = 3.3 \text{ kJ} = 3.3 \times 10^3 \text{ J}$

$\Delta S° = [89 - 158.2] \text{ J/K} = -69.2 \text{ J/K}$

$\Delta G° = \Delta H° - T\Delta S°$

$\Delta G° = 3.3 \times 10^3 \text{ J} - (298 \text{ K})(-69.2 \text{ J/K}) = 23.921 \times 10^3 \text{ J}$

Now substitute numerical values into the equation relating $\ln K$ and $\Delta G°$.

$$\ln K = \frac{\Delta G°}{-RT} = \frac{23.921 \times 10^3}{-8.31 \times 298} = -9.6599$$

$$K = e^{-9.6599} = 6.37 \times 10^{-5} = 6 \times 10^{-5}$$

b. The change in entropy is negative, greater order, so this causes H_3PO_4 to be a weak acid. The enthalpy change hinders the acid strength of H_3PO_4, and the entropy is a very important term.

■ Solutions to Cumulative-Skills Problems

18.69 For the dissociation of HBr, assume ΔH and ΔS are constant over the temperature range from 25°C to 375°C, and calculate the value of each to use to calculate K at 375°C. Start by calculating $\Delta H°$ and $\Delta S°$ at 25°C, using $\Delta H_f°$ and $S°$ values.

	2HBr(g)	$\rightarrow$	H$_2$(g)	+	Br$_2$(g)
$\Delta H_f°$:	2 x (-36.44)		0		30.91 kJ
$S°$:	2 x 198.6		130.6		245.3 J/K

Calculate $\Delta H°$ and $\Delta S°$ from these values.

$$\Delta H° = [30.91 - 2(-36.44)] \text{ kJ} = 103.79 \text{ kJ}$$

$$\Delta S° = [245.3 + 130.6 - 2(198.6)] \text{ J/K} = -21.3 \text{ J/K}$$

Substitute $\Delta H°$, $\Delta S°$ (= -0.02130 kJ/K), and T (648 K) into the equation for $\Delta G_T°$.

$$\Delta G_T° = \Delta H° - T\Delta S° = 103.79 \text{ kJ} - (648 \text{ K})(-0.02130 \text{ kJ/K}) = 117.\underline{5}9 \text{ kJ}$$

$$= 117.\underline{5}9 \times 10^3 \text{ J}$$

Now substitute numerical values into the equation relating ln K and $\Delta G°$ (= $\Delta G_T°$).

$$\ln K = \frac{\Delta G°}{-RT} = \frac{117.59 \times 10^3}{-8.31 \times 648} = -21.\underline{8}37$$

$$K = e^{-21.837} = \underline{3}.28 \times 10^{-10}$$

Assuming x equals [H$_2$] equals [Br$_2$], and assuming [HBr] = (1.00 - 2x) $\cong$ 1.00 atm, substitute into the equilibrium expression:

$$K = \frac{[\text{H}_2][\text{Br}_2]}{[\text{HBr}]^2} = \frac{(x)(x)}{(1.00)^2} = \underline{3}.28 \times 10^{-10}$$

Solve for the approximate pressure of x:

$$x = \sqrt{(3.28 \times 10^{-10})(1.00)^2} = \underline{1}.81 \times 10^{-5} \text{ atm}$$

(continued)

The percent dissociation at 1.00 atm is

$$\text{Percent dissociation} = \frac{2\,(1.81 \times 10^{-5}\ \text{atm})}{1.00\ \text{atm}} \times 100\% = 0.00\underline{3}6\ \text{percent}$$

Based on Le Châtelier's principle, pressure has no effect on equilibrium. Therefore the percent dissociation is 0.004 percent at 1.00 atm and at 10.0 atm.

18.71 First calculate $\Delta G°$ at each temperature, using $\Delta G° = -RT \ln K$:

25.0°C: $\Delta G° = -(0.008314\ \text{kJ/K})(298.2\ \text{K})(\ln 1.754 \times 10^{-5}) = 27.15\underline{5}1\ \text{kJ}$

50.0°C: $\Delta G°_T = -(0.008314\ \text{kJ/K})(323.2\ \text{K})(\ln 1.633 \times 10^{-5}) = 29.62\underline{3}7\ \text{kJ}$

Next solve two equations in two unknowns assuming $\Delta H°$ and $\Delta S°$ are constant over the range of 25.0°C to 50.0°C. Use 0.2982 K(kJ/J) and 0.3232 K(kJ/J) to convert $\Delta S°$ in J to $T\Delta S°$ in kJ.

1. 27.1551 kJ $= \Delta H° - [0.2982\ \text{K(kJ/J)}\ \Delta S°]$

2. 29.6237 kJ $= \Delta H° - [0.3232\ \text{K(kJ/J)}\ \Delta S°]$

Then rearrange equation 2, and substitute for $\Delta H°$ into equation 2:

3a. $\Delta H° = 0.3232\ \text{K(kJ/J)}\ \Delta S° + 29.6237\ \text{kJ}$

3b. 27.1551 kJ $= [0.3232\ \text{K(kJ/J)}\ \Delta S° + 29.6237\ \text{kJ}] - [0.2982\ \text{K(kJ/J)}\ \Delta S°]$

Solve for $\Delta S°$:

$$\Delta S° = \frac{(29.6237 - 27.1551)\ \text{kJ}}{(0.2982 - 0.3232)\ \text{K/(kJ/J)}} = -98.\underline{7}4 = -98.7\ \text{J/K}$$

Substitute this value into equation 3a and solve for $\Delta H°$:

$\Delta H° = [(0.3232)\ \text{K(kJ/J)} \times (-98.74\ \text{J/K})] + 29.6237\ \text{kJ} = -2.2\underline{8}9 = -2.29\ \text{kJ}$

19. ELECTROCHEMISTRY

■ Answers to Review Questions

19.1   A voltaic cell is an electrochemical cell in which a spontaneous reaction generates an electric current (energy). An electrolytic cell is an electrochemical cell that requires electrical current (energy) to drive a nonspontaneous reaction to the right.

19.2   The SI unit of electrical potential is the volt (V).

19.3   It is necessary to measure the voltage of a voltaic cell when no current is flowing because the cell voltage exhibits its maximum value only when no current flows. Even if the current flows just for the time of measurement, the voltage drops enough so that what is measured is significantly less than the maximum.

19.4   Standard electrode potentials are defined relative to a standard electrode potential of zero volts (0.00, 0.000 V, etc.) for the $H^+/H_2(g)$ electrode. Because the cell emf is measured using the hydrogen electrode at standard conditions and a second electrode at standard conditions, the cell emf equals the $E°$ of the half-reaction at the second electrode.

19.5   The mathematical relationships are as follows:

$$\Delta G° = -nFE°_{cell}$$

$$\Delta G° = -RT \ln K$$

Combining these two equations gives

$$\ln K = \frac{nFE°_{cell}}{RT}$$

19.6   The zinc-carbon cell has a zinc can as the anode; the cathode is a graphite rod surrounded by a paste of manganese dioxide and carbon black. Around this is a second paste of ammonium and zinc chlorides. The electrode reactions involve oxidation of zinc metal to zinc(II) ion and reduction of $MnO_2(s)$ to $Mn_2O_3(s)$ at the cathode. The lead storage battery consists of a spongy lead anode and a lead dioxide cathode, both immersed in aqueous sulfuric acid. At the anode, the lead is oxidized to lead sulfate; at the cathode, lead dioxide is reduced to lead sulfate.

19.7   During the rusting of iron, one end of a drop of water exposed to air acts as one electrode of a voltaic cell; at this electrode, an oxygen molecule is reduced by four electrons to four hydroxide ions. Oxidation of metallic iron to iron(II) ion at the center of the drop of water supplies the electrons, and the center serves as the other electrode of the voltaic cell. Thus electrons flow from the center of the drop through the iron to the end of the drop.

19.8   When iron or steel is connected to an active metal such as zinc, a voltaic cell is formed with zinc as the anode and iron as the cathode. Any type of moisture forms the electrolyte solution, and the zinc metal is then oxidized to zinc(II) ion in preference to the oxidation of iron metal. Oxygen is reduced at the cathode to hydroxide ions. If iron or steel is exposed to oxygen while connected to a less active metal such as tin, a voltaic cell is formed with iron as the anode and tin as the cathode, and iron is oxidized to iron(II) ion rather than tin being oxidized to tin(II) ion. Thus exposed iron corrodes rapidly in a tin can. Fortunately, as long as the iron is covered by the tin, it cannot corrode.

19.9   The addition of an ionic species such as strongly ionized sulfuric acid facilitates the passage of current through the solution.

19.10   Sodium metal can be prepared by electrolysis of molten sodium chloride.

19.11   The anode reaction in the electrolysis of molten potassium hydroxide is

$$4OH^- \rightarrow O_2(g) + 2H_2O(g) + 4e^-$$

19.12   The reason different products are obtained is that water instead of Na^+ is reduced at the cathode during the electrolysis of aqueous NaCl. This is because water has a more positive $E°$ (smaller decomposition voltage). At the anode, water instead of chloride ion is oxidized because water has a less positive $E°$ (smaller decomposition voltage).

■ Solutions to Practice Problems

Note on significant figures: If the final answer to a solution needs to be rounded off, it is given first with one nonsignificant figure, and the last significant figure is underlined. The final answer is then rounded to the correct number of significant figures. In multiple-step problems, intermediate answers are given with at least one nonsignificant figure; however, only the final answer has been rounded off.

19.19   In balancing oxidation-reduction reactions in acid, the four steps in the text will be followed. For part a, each step is shown. For the other parts, only a summary is shown.

   a. Assign oxidation numbers to the skeleton equation (Step 1).

$$\overset{+6}{Cr_2O_7{}^{2-}} + \overset{+3}{C_2O_4{}^{2-}} \rightarrow \overset{+3}{Cr^{3+}} + \overset{+4}{CO_2}$$

Separate into two incomplete half-reactions (Step 2). Note that carbon is oxidized (increases in oxidation number), and chromium is reduced (decreases in oxidation number).

$$C_2O_4{}^{2-} \rightarrow CO_2$$
$$Cr_2O_7{}^{2-} \rightarrow Cr^{3+}$$

Balance each half-reaction separately. The oxidation half-reaction is not balanced in C, so place a two in front of CO_2 (Step 3a). Finally, add two electrons to the right side to balance the charge (Step 3d). The balanced oxidation half-reaction is

$$C_2O_4{}^{2-} \rightarrow 2CO_2 + 2e^-$$

The reduction half-reaction is not balanced in Cr, so place a two in front of Cr^{3+} (Step 3a). Add seven H_2O to the right side to balance O atoms (Step 3b), and add fourteen H^+ ion to the left side to balance H atoms (step 3c). Finally, add six electrons to the left side to balance the charge (Step 3d). The balanced reduction half-reaction is

$$Cr_2O_7{}^{2-} + 14H^+ + 6e^- \rightarrow 2Cr^{3+} + 7H_2O$$

Multiply the oxidation half-reaction by three so that the electrons cancel when added (Step 4a).

$$3C_2O_4{}^{2-} \rightarrow 6CO_2 + 6e^-$$
$$Cr_2O_7{}^{2-} + 14H^+ + 6e^- \rightarrow 2Cr^{3+} + 7H_2O$$
$$\overline{Cr_2O_7{}^{2-} + 3C_2O_4{}^{2-} + 14H^+ + \cancel{6e^-} \rightarrow 2Cr^{3+} + 6CO_2 + 7H_2O + \cancel{6e^-}}$$

(continued)

The equation does not need to be simplified any further (Step 4b). The net ionic equation is

$$Cr_2O_7^{2-} + 3C_2O_4^{2-} + 14H^+ \rightarrow 2Cr^{3+} + 6CO_2 + 7H_2O$$

b. The two balanced half-reactions are

$$Cu \rightarrow Cu^{2+} + 2e^- \qquad \text{(oxidation)}$$
$$NO_3^- + 4H^+ + 3e^- \rightarrow NO + 2H_2O \qquad \text{(reduction)}$$

Multiply the oxidation half-reaction by three and the reduction half-reaction by four, and then add together. Cancel the six electrons from each side. No further simplification is needed. The balanced equation is

$$3Cu + 2NO_3^- + 8H^+ \rightarrow 3Cu^{2+} + 2NO + 4H_2O$$

c. The two balanced half-reactions are

$$HNO_2 + H_2O \rightarrow NO_3^- + 3H^+ + 2e^- \qquad \text{(oxidation)}$$
$$MnO_2 + 4H^+ + 2e^- \rightarrow Mn^{2+} + 2H_2O \qquad \text{(reduction)}$$

Add the two half-reactions together, and cancel the two electrons from each side. Also cancel three H^+ and one H_2O from each side. The balanced equation is

$$MnO_2 + HNO_2 + H^+ \rightarrow Mn^{2+} + NO_3^- + H_2O$$

d. The two balanced half-reactions are

$$Mn^{2+} + 4H_2O \rightarrow MnO_4^- + 8H^+ + 5e^- \qquad \text{(oxidation)}$$
$$PbO_2 + SO_4^{2-} + 4H^+ + 2e^- \rightarrow PbSO_4 + 2H_2O \qquad \text{(reduction)}$$

Multiply the oxidation half-reaction by two and the reduction half-reaction by five, and then add together. Cancel the ten electrons from each side. Also cancel sixteen H^+ and eight H_2O from each side. The balanced equation is

$$5PbO_2 + 2Mn^{2+} + 5SO_4^{2-} + 4H^+ \rightarrow 5PbSO_4 + 2MnO_4^- + 2H_2O$$

19.21 When balancing an oxidation-reduction reaction in basic solution, it will first be balanced as if it were in acidic solution; then the extra two steps in the text will be followed.

a. The two balanced half-reactions are

$$Mn^{2+} + 2H_2O \rightarrow MnO_2 + 4H^+ + 2e^- \quad \text{(oxidation)}$$

$$H_2O_2 + 2H^+ + 2e^- \rightarrow 2H_2O \quad \text{(reduction)}$$

Add the two half-reactions together, and cancel the two electrons from each side. Also cancel two H^+ and two H_2O from each side. The balanced equation in acidic solution is

$$Mn^{2+} + H_2O_2 \rightarrow MnO_2 + 2H^+$$

Now add two OH^- to each side (Step 5). Simplify by combining the H^+ and OH^- to give H_2O. No further simplification is required (Step 6). The balanced equation in basic solution is

$$Mn^{2+} + H_2O_2 + 2OH^- \rightarrow MnO_2 + 2H_2O$$

b. The two balanced half-reactions are

$$NO_2^- + H_2O \rightarrow NO_3^- + 2H^+ + 2e^- \quad \text{(oxidation)}$$

$$MnO_4^- + 4H^+ + 3e^- \rightarrow MnO_2 + 2H_2O \quad \text{(reduction)}$$

Multiply the oxidation half-reaction by three and the reduction half-reaction by two, and then add together. Cancel the six electrons from each side. Also cancel twelve H^+ and six H_2O from each side. The balanced equation in acidic solution is

$$2MnO_4^- + 3NO_2^- + 2H^+ \rightarrow 2MnO_2 + 3NO_3^- + H_2O$$

Now add two OH^- to each side. Simplify by combining the H^+ and OH^- to give H_2O. Then cancel one H_2O from each side. The balanced equation in basic solution is

$$2MnO_4^- + 3NO_2^- + H_2O \rightarrow 2MnO_2 + 3NO_3^- + 2OH^-$$

c. The two balanced half-reactions are

$$Mn^{2+} + 2H_2O \rightarrow MnO_2 + 4H^+ + 2e^- \quad \text{(oxidation)}$$

$$ClO_3^- + 2H^+ + e^- \rightarrow ClO_2 + H_2O \quad \text{(reduction)}$$

(continued)

Multiply the reduction half-reaction by two, and then add together. Cancel the two electrons from each side. Also cancel four H^+ and two H_2O from each side. The balanced equation in acidic solution is

$$Mn^{2+} + 2ClO_3^- \rightarrow MnO_2 + 2ClO_2$$

Since there are no H^+ on either side, no further simplification is needed. The balanced equation in basic solution is identical to the balanced equation in acidic solution.

d. The two balanced half-reactions are

$$NO_2 + H_2O \rightarrow NO_3^- + 2H^+ + e^- \qquad \text{(oxidation)}$$
$$MnO_4^- + 4H^+ + 3e^- \rightarrow MnO_2 + 2H_2O \qquad \text{(reduction)}$$

Multiply the oxidation half-reaction by three, and then add together. Cancel the three electrons from each side. Also cancel four H^+ and two H_2O from each side. The balanced equation in acidic solution is

$$MnO_4^- + 3NO_2 + H_2O \rightarrow MnO_2 + 3NO_3^- + 2H^+$$

Now add two OH^- to each side. Simplify by combining the H^+ and OH^- to give H_2O. Then cancel one H_2O from each side. The balanced equation in basic solution is

$$MnO_4^- + 3NO_2 + 2OH^- \rightarrow MnO_2 + 3NO_3^- + H_2O$$

19.23 a. The two balanced half-reactions are

$$8H_2S \rightarrow S_8 + 16H^+ + 16e^- \qquad \text{(oxidation)}$$
$$NO_3^- + 2H^+ + e^- \rightarrow NO_2 + H_2O \qquad \text{(reduction)}$$

Multiply the reduction half-reaction by sixteen, and then add together. Cancel the sixteen electrons and sixteen H^+ from each side. The balanced equation is

$$8H_2S + 16NO_3^- + 16H^+ \rightarrow S_8 + 16NO_2 + 16H_2O$$

b. The two balanced half-reactions are

$$Cu \rightarrow Cu^{2+} + 2e^- \qquad \text{(oxidation)}$$
$$NO_3^- + 4H^+ + 3e^- \rightarrow NO + 2H_2O \qquad \text{(reduction)}$$

Multiply the oxidation half-reaction by three and the reduction half-reaction by two, and then add together. Cancel the six electrons from each side. The balanced equation is

$$2NO_3^- + 3Cu + 8H^+ \rightarrow 2NO + 3Cu^{2+} + 4H_2O$$

(continued)

c. The two balanced half-reactions are

$$SO_2 + 2H_2O \rightarrow SO_4^{2-} + 4H^+ + 2e^- \qquad \text{(oxidation)}$$

$$MnO_4^- + 8H^+ + 5e^- \rightarrow Mn^{2+} + 4H_2O \qquad \text{(reduction)}$$

Multiply the oxidation half-reaction by five and the reduction half-reaction by two, and then add together. Cancel the ten electrons from each side. Also cancel sixteen H^+ and eight H_2O from each side. The balanced equation is

$$2MnO_4^- + 5SO_2 + 2H_2O \rightarrow 5SO_4^{2-} + 2Mn^{2+} + 4H^+$$

d. The two balanced half-reactions are

$$Sn(OH)_3^- + 3H_2O \rightarrow Sn(OH)_6^{2-} + 3H^+ + 2e^- \qquad \text{(oxidation)}$$

$$Bi(OH)_3 + 3H^+ + 3e^- \rightarrow Bi + 3H_2O \qquad \text{(reduction)}$$

Multiply the oxidation half-reaction by three and the reduction half-reaction by two, and then add together. Cancel the six electrons from each side. Also cancel six H^+ and six H_2O from each side. The balanced equation in acidic solution is

$$2Bi(OH)_3 + 3Sn(OH)_3^- + 3H_2O \rightarrow 3Sn(OH)_6^{2-} + 2Bi + 3H^+$$

Now add three OH^- to each side. Simplify by combining the H^+ and OH^- to give H_2O. Then cancel three H_2O on each side. The balanced equation in basic solution is

$$2Bi(OH)_3 + 3Sn(OH)_3^- + 3OH^- \rightarrow 3Sn(OH)_6^{2-} + 2Bi$$

19.25 Sketch of the cell:

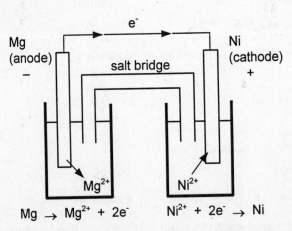

19.27 The electrode half-reactions and the overall cell reaction are

Anode: $Zn(s) + 2OH^-(aq) \rightarrow Zn(OH)_2(s) + 2e^-$

Cathode: $Ag_2O(s) + H_2O\ (l) + 2e^- \rightarrow 2Ag(s) + 2OH^-(aq)$

Overall: $Zn(s) + Ag_2O(s) + H_2O(l) \rightarrow Zn(OH)_2(s) + 2Ag(s)$

19.29 Because of its less negative $E°$, Pb^{2+} is reduced at the cathode and is written on the right; $Sn(s)$ is oxidized at the anode and is written first, at the left, in the cell notation. The notation is $Sn(s)|Sn^{2+}(aq)||Pb^{2+}(aq)|Pb(s)$.

19.31 Because of its less negative $E°$, H^+ is reduced at the cathode and is written on the right; $Ni(s)$ is oxidized at the anode and is written first, at the left, in the cell notation. The notation is $Ni(s)|Ni^{2+}(1\ M)||H^+(1\ M)|H_2(g)|Pt$.

19.33 The half-cell reactions, the overall cell reaction, and the sketch are

$$Cd(s) \rightarrow Cd^{2+}(aq) + 2e^-$$
$$Ni^{2+}(aq) + 2e^- \rightarrow Ni(s)$$
$$\overline{Cd(s) + Ni^{2+}(aq) \rightarrow Cd^{2+}(aq) + Ni(s)}$$

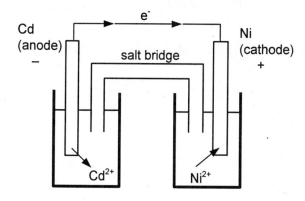

19.35   The half-cell reactions are

$$2Fe^{3+}(aq) + 2e^- \rightarrow 2Fe^{2+}(aq)$$
$$Zn(s) \rightarrow Zn^{2+}(aq) + 2e^-$$

n equals two, and the maximum work for the reaction as written is

$$w_{max} = -nFE_{cell} = -2 \times 9.65 \times 10^4 \, C \times 0.72 \, V = -1.\underline{3}89 \times 10^5 \, C{\bullet}V$$

$$= -1.\underline{3}89 \times 10^5 \, J$$

Because this is the work obtained by reduction of one mol of Zn, the work for 35.0 g of Zn is

$$35.0 \text{ g Zn} \times \frac{1 \text{ mol Zn}}{65.39 \text{ g Zn}} \times \frac{-1.389 \times 10^5 \, J}{1 \text{ mol Zn}} = -7.4\underline{3}8 \times 10^4 = = -7.44 \times 10^4 \, J$$

19.37   The half-reactions and corresponding electrode potentials are as follows

$NO_3^-(aq) + 4H^+(aq) + 3e^- \rightarrow NO(g) + 2H_2O(l)$	0.96 V
$O_2(g) + 4H^+(aq) + 4e^- \rightarrow 2H_2O(l)$	1.23 V
$MnO_4^-(aq) + 8H^+(aq) + 5e^- \rightarrow Mn^{2+}(aq) + 4H_2O(l)$	1.49 V

The order by increasing oxidizing strength is $NO_3^-(aq)$, $O_2(g)$, $MnO_4^-(aq)$.

19.39   The half-reactions and corresponding electrode potentials are as follows

$Zn^{2+}(aq) + 2e^- \rightarrow Zn(s)$	-0.76 V
$Fe^{2+}(aq) + 2e^- \rightarrow Fe(s)$	-0.41 V
$Cu^{2+}(aq) + e^- \rightarrow Cu^+(aq)$	0.16 V

Zn(s) is the strongest, and $Cu^+(aq)$ is the weakest.

19.41 a. In this reaction, Sn^{4+} is the oxidizing agent on the left side; Fe^{3+} is the oxidizing reagent on the right side. The corresponding standard electrode potentials are

$Sn^{4+}(aq) + 2e^- \rightarrow Sn^{2+}(aq)$	$E° = 0.15 \, V$
$Fe^{3+}(aq) + e^- \rightarrow Fe^{2+}(aq)$	$E° = 0.77 \, V$

The stronger oxidizing agent is the one involved in the half-reaction with the more positive standard electrode potential, so Fe^{3+} is the stronger oxidizing agent. The reaction is nonspontaneous as written.

(continued)

b. In this reaction, MnO_4^- is the oxidizing agent on the left side; O_2 is the oxidizing reagent on the right side. The corresponding standard electrode potentials are

$$O_2(g) + 4H^+(aq) + 4e^- \rightarrow 2H_2O(l) \qquad\qquad E^\circ = 1.23 \text{ V}$$

$$MnO_4^-(aq) + 8H^+(aq) + 5e^- \rightarrow Mn^{2+}(aq) + 4\,H_2O(l) \qquad E^\circ = 1.49 \text{ V}$$

The stronger oxidizing agent is the one involved in the half-reaction with the more positive standard electrode potential, so MnO_4^- is the stronger oxidizing agent. The reaction is spontaneous as written.

19.43   The reduction half-reactions and standard electrode potentials are

$$Br_2(l) + 2e^- \rightarrow 2Br^-(aq) \qquad\qquad E^\circ = 1.07 \text{ V}$$

$$Cl_2(g) + 2e^- \rightarrow 2Cl^-(aq) \qquad\qquad E^\circ = 1.36 \text{ V}$$

$$F_2(g) + 2e^- \rightarrow 2F^-(aq) \qquad\qquad E^\circ = 2.87 \text{ V}$$

From these, you see the order of increasing oxidizing strength is Br_2, Cl_2, F_2. Therefore chlorine gas will oxidize Br^- but will not oxidize F^-. The balanced equation for the reaction is

$$Cl_2(g) + 2Br^-(aq) \rightarrow 2Cl^-(aq) + Br_2(l)$$

19.45   The half-reactions and standard electrode potentials are

$$Al(s) \rightarrow Al^{3+}(aq) + 3e^- \qquad\qquad -E^\circ_{Al} = 1.66 \text{ V}$$

$$Hg_2^{2+}(aq) + 2e^- \rightarrow 2Hg(l) \qquad\qquad E^\circ_{Hg} = 0.80 \text{ V}$$

Obtain the cell emf by adding the half-cell potentials.

$$E^\circ_{cell} = E^\circ_{Hg} - E^\circ_{Al} = 0.80 \text{ V} + 1.66 \text{ V} = 2.46 \text{ V}$$

19.47   The half-cell reactions, the corresponding half-cell potentials, and their sums are displayed below:

$$3Cu(s) \rightarrow 3Cu^{2+}(aq) + 6e^- \qquad -E^\circ = -0.34 \text{ V}$$

$$2NO_3^-(aq) + 8H^+(aq) + 6e^- \rightarrow 2NO(g) + 4H_2O(l) \qquad E^\circ = 0.96 \text{ V}$$

$$3Cu(s) + 2NO_3^-(aq) + 8H^+(aq) \rightarrow$$

$$3Cu^{2+}(aq) + 2NO(g) + 4H_2O(l) \qquad E^\circ_{cell} = 0.62 \text{ V}$$

(continued)

Note that each half-reaction involves six electrons, hence $n = 6$. Therefore,

$$\Delta G° = -nFE°_{cell} = -6 \times 9.65 \times 10^4 \text{ C} \times 0.62 \text{ V} = -3.\underline{5}8 \times 10^5 \text{ J} = -3.6 \times 10^5 \text{ J}$$

Thus the standard free-energy change is -3.6×10^2 kJ.

19.49   Write the equation with $\Delta G_f°$'s beneath each substance.

	Mg(s)	+	2Ag$^+$(aq)	→	Mg^{2+}(aq)	+	2Ag(s)
$\Delta G_f°$:	0		2 x 77.111		-456.01		0 kJ

Hence,

$$\Delta G° = \Sigma n \Delta G_f°(\text{products}) - \Sigma m \Delta G_f°(\text{reactants})$$

$$= [-456.01 - 2 \times 77.111] \text{ kJ} = -610.\underline{2}3 \text{ kJ} = -6.10\underline{2}3 \times 10^5 \text{ J}$$

Obtain n by splitting the reaction into half-reactions.

$$\text{Mg}(s) \rightarrow \text{Mg}^{2+}(aq) + 2e^-$$
$$2\text{Ag}^+(aq) + 2e^- \rightarrow 2\text{Ag}(s)$$

Each half-reaction involves two electrons, so $n = 2$. Therefore,

$$\Delta G° = -nFE°_{cell}$$

$$-6.1023 \times 10^5 \text{ J} = -2 \times 9.65 \times 10^4 \text{ C} \times E°_{cell}$$

Rearrange and solve for $E°_{cell}$. Recall that $J = C \cdot V$.

$$E°_{cell} = \frac{-6.1023 \times 10^5 \text{ J}}{-2 \times 9.65 \times 10^4 \text{ C}} = 3.1\underline{6}18 = 3.16 \text{ V}$$

19.51   The half-reactions and standard electrode potentials are

$$\text{Pb}(s) \rightarrow \text{Pb}^{2+}(aq) + 2e^- \qquad -E°_{Pb} = 0.13 \text{ V}$$
$$2\text{Fe}^{3+}(aq) + 2e^- \rightarrow 2\text{Fe}^{2+}(aq) \qquad E°_{Fe^{3+}} = 0.77 \text{ V}$$

The standard emf for the cell is

$$E°_{cell} = E°_{Fe^{3+}} - E°_{Pb} = 0.77 \text{ V} + 0.13 \text{ V} = 0.90 \text{ V}$$

(continued)

Note that $n = 2$. Substitute into the equation relating $E°$ and K. Note that $K = K_c$.

$$0.90 \text{ V} = \frac{0.0592}{2} \log K_c$$

Solving for K_c, you get

$$\log K_c = 30.40$$

Take the antilog of both sides:

$$K_c = \text{antilog } (30.40) = 2.5 \times 10^{30} = 10^{30}$$

19.53   The half-cell reactions, the corresponding half-cell potentials, and their sums are displayed below:

$2Cr(s) \rightarrow 2Cr^{3+}(aq) + 6e^-$	$-E° = 0.74$ V	
$3Ni^{2+}(aq) + 6e^- \rightarrow 3Ni(s)$	$E° = -0.23$ V	
$2Cr(s) + 3Ni^{2+}(aq) \rightarrow 2Cr^{3+}(aq) + 3Ni(s)$	$E°_{cell} = 0.51$ V	

Note that n equals six. The reaction quotient is

$$Q = \frac{[Cr^{3+}]^2}{[Ni^{2+}]^3} = \frac{(1.0 \times 10^{-3})^2}{(1.5)^3} = 2.962 \times 10^{-7}$$

The standard emf is 0.51 V, so the Nernst equation becomes

$$E_{cell} = E°_{cell} - \frac{0.0592}{n} \log Q$$

$$= 0.51 - \frac{0.0592}{6} \log (2.962 \times 10^{-7})$$

$$= 0.51 - (-0.06441) = 0.574 = 0.57 \text{ V}$$

19.55   The overall reaction is

$$Cd(s) + Ni^{2+}(aq) \rightarrow Cd^{2+}(aq) + Ni(s)$$

Note that n equals two. The reaction quotient is

$$Q = \frac{[Cd^{2+}]}{[Ni^{2+}]} = \frac{[Cd^{2+}]}{1.0} = [Cd^+]$$

The standard emf is 0.170 V, and the emf is 0.240 V, so the Nernst equation becomes

$$E_{cell} = E^\circ_{cell} - \frac{0.0592}{n}\log Q$$

$$0.240 = 0.170 - \frac{0.0592}{2}\log Q$$

Rearrange and solve for log Q

$$\log Q = \frac{2}{0.0592} \times (0.240 - 0.170) = -2.\underline{3}648$$

Take the antilog of both sides

$$Q = [Cd^{2+}] = antilog\,(-2.3648) = \underline{4}.31 \times 10^{-3} = 0.004\ M$$

The Cd^{2+} concentration is 0.004 M.

19.57 a. The species you should consider for half-reactions are Na^+, SO_4^{2-}, and H_2O. The possible cathode reactions are

$$Na^+(aq) + e^- \rightarrow Na(s) \qquad\qquad E^\circ = -2.71\ V$$
$$2H_2O(l) + 2e^- \rightarrow H_2(g) + 2OH^-(aq) \qquad E^\circ = -0.83\ V$$

Because the electrode potential for H_2O is larger (less negative), it is easier to reduce.

The possible anode reactions are

$$2H_2O(l) \rightarrow O_2(g) + 4H^+(aq) + 4e^- \qquad E^\circ = -1.23\ V$$
$$2SO_4^{2-}(aq) \rightarrow S_2O_4^{2-}(aq) + 2e^- \qquad E^\circ = -2.01\ V$$

Because the electrode potential for H_2O is less negative, it is easier to oxidize.

(continued)

The expected half-reactions are

$2H_2O(l) \rightarrow O_2(g) + 4H^+(aq) + 4e^-$ $\qquad$ $E° = -1.23$ V

$4H_2O(l) + 4e^- \rightarrow 2H_2(g) + 4OH^-(aq)$ $\qquad$ $E° = -0.83$ V

The overall reaction is

$2H_2O(l) \rightarrow 2H_2(g) + O_2(g)$

b. The species you should consider for half-reactions are K^+, Br^-, and H_2O. The possible cathode reactions are

$K^+(aq) + e^- \rightarrow K(s)$ $\qquad$ $E° = -2.92$ V

$2H_2O(l) + 2e^- \rightarrow H_2(g) + 2OH^-(aq)$ $\qquad$ $E° = -0.83$ V

Because the electrode potential for H_2O is larger (less negative), it is easier to reduce. The possible anode reactions are

$2H_2O(l) \rightarrow O_2(g) + 4H^+(aq) + 4e^-$ $\qquad$ $E° = -1.23$ V

$2Br^-(aq) \rightarrow Br_2(l) + 2e^-$ $\qquad$ $E° = -1.07$ V

Because the electrode potential for Br^- is less negative, it is easier to oxidize.

The expected half-reactions are

$2Br^-(aq) \rightarrow Br_2(l) + 2e^-$ $\qquad$ $E° = -1.07$ V

$2H_2O(l) + 2e^- \rightarrow H_2(g) + 2OH^-(aq)$ $\qquad$ $E° = -0.83$ V

The overall reaction is

$2Br^-(aq) + 2H_2O(l) \rightarrow Br_2(l) + H_2(g) + 2OH^-$

19.59 The conversion of grams of aluminum (2.78 kg = 2.78×10^3 g) to coulombs required to give this amount of aluminum is

$$2.78 \times 10^3 \text{ g} \times \frac{1 \text{ mol Al}}{26.98 \text{ g}} \times \frac{3 \text{ mol e}^-}{1 \text{ mol Al}} \times \frac{9.65 \times 10^4 \text{ C}}{1 \text{ mol e}^-}$$

$$= 2.9\underline{8}2 \times 10^7 = 2.98 \times 10^7 \text{ C}$$

19.61   The conversion of coulombs to grams of lithium is

$$5.00 \times 10^3 \text{ C} \times \frac{1 \text{ mol e}^-}{9.65 \times 10^4 \text{ C}} \times \frac{1 \text{ mol Li}}{1 \text{ mol e}^-} \times \frac{6.941 \text{ g Li}}{1 \text{ mol Li}} = 0.35\underline{9}6$$

$$= 0.360 \text{ g Li}$$

■ Solutions to General Problems

19.63 a. The two balanced half-reactions are

$$8S^{2-} \rightarrow S_8 + 16e^- \qquad \text{(oxidation)}$$
$$MnO_4^- + 4H^+ + 3e^- \rightarrow MnO_2 + 2H_2O \qquad \text{(reduction)}$$

Multiply the oxidation half-reaction by three and the reduction half-reaction by sixteen, and then add together. Cancel the forty-eight electrons from each side. The balanced equation in acidic solution is

$$16MnO_4^- + 24S^{2-} + 64H^+ \rightarrow 16MnO_2 + 3S_8 + 32H_2O$$

Now add sixty-four OH^- to each side. Simplify by combining the H^+ and OH^- to give H_2O. Then cancel thirty-two H_2O on each side. The balanced equation in basic solution is

$$16MnO_4^-(aq) + 24S^{2-}(aq) + 32H_2O(l) \rightarrow 16MnO_2(aq) + 3S_8(aq) + 64OH^-(l)$$

b. The two balanced half-reactions are

$$HSO_3^- + H_2O \rightarrow SO_4^{2-} + 3H^+ + 2e^- \qquad \text{(oxidation)}$$
$$IO_3^- + 6H^+ + 6e^- \rightarrow I^- + 3H_2O \qquad \text{(reduction)}$$

Multiply the oxidation half-reaction by three, and then add together. Cancel the six electrons from each side. Also cancel six H^+ and three H_2O from each side. The balanced equation is

$$IO_3^-(aq) + 3HSO_3^-(aq) \rightarrow I^- + 3SO_4^{2-}(aq) + 3H^+(aq)$$

(continued)

c. The two balanced half-reactions are

$$Fe(OH)_2 + H_2O \rightarrow Fe(OH)_3 + H^+ + e^- \qquad \text{(oxidation)}$$
$$CrO_4^{2-} + 4H^+ + 3e^- \rightarrow Cr(OH)_4^- \qquad \text{(reduction)}$$

Multiply the oxidation half-reaction by three, and then add together. Cancel the three electrons, and cancel three H^+ from each side. The balanced equation in acidic solution is

$$CrO_4^{2-}(aq) + 3Fe(OH)_2(s) + H^+(aq) + 3H_2O(l) \rightarrow Cr(OH)_4^-(aq) + 3Fe(OH)_3(s)$$

Now add one OH^- to each side. Simplify by combining the H^+ and OH^- to give H_2O. Then combine the H_2Os on the left side into a single term. The balanced equation in basic solution is

$$CrO_4^{2-}(aq) + 3Fe(OH)_2(aq) + 4H_2O(l) \rightarrow Cr(OH)_4^-(aq) + 3Fe(OH)_3(s) + OH^-(aq)$$

19.65 This reaction takes place in basic solution. The skeleton equation is

$$Fe(OH)_2(s) + O_2(g) \rightarrow Fe(OH)_3(s)$$

The two balanced half-reactions are

$$Fe(OH)_2 + H_2O \rightarrow Fe(OH)_3 + H^+ + e^- \qquad \text{(oxidation)}$$
$$O_2 + 4H^+ + 4e^- \rightarrow 2H_2O \qquad \text{(reduction)}$$

Multiply the oxidation half-reaction by four, and then add together. Cancel the four electrons from each side. Also cancel four H^+ and two H_2O from each side. The balanced equation in acidic or in basic solution is

$$4Fe(OH)_2(s) + O_2(g) + 2H_2O(l) \rightarrow 4Fe(OH)_3(s)$$

19.67 The cell notation is $Ca(s)|Ca^{2+}(aq)||Cl^-(aq)|Cl_2(g)|Pt(s)$. The reactions are

Anode: $Ca(s) \rightarrow Ca^{2+}(aq) + 2e^- \qquad -E° = 2.76 \text{ V}$
Cathode: $Cl_2(g) + 2e^- \rightarrow 2Cl^-(aq) \qquad E° = 1.36 \text{ V}$

$$E°_{cell} = 1.36 \text{ V} + 2.76 \text{ V} = 4.12 \text{ V}$$

19.69 a. In this reaction, Fe^{3+} is the oxidizing agent on the left side; Ni^{2+} is the oxidizing reagent on the right side. The corresponding standard electrode potentials are

$$Ni^{2+}(aq) + 2e^- \rightarrow Ni(s) \qquad\qquad E^\circ = -0.23 \text{ V}$$

$$Fe^{3+}(aq) + e^- \rightarrow Fe^{2+}(aq) \qquad\qquad E^\circ = 0.77 \text{ V}$$

The stronger oxidizing agent is the one involved in the half-reaction with the more positive standard electrode potential, so Fe^{3+} is the stronger oxidizing agent. The oxidation of nickel by iron(III) is a spontaneous reaction.

b. In this reaction, Fe^{3+} is the oxidizing agent on the left side; Sn^{4+} is the oxidizing reagent on the right side. The corresponding standard electrode potentials are

$$Sn^{4+}(aq) + 2e^- \rightarrow Sn^{2+}(aq) \qquad\qquad E^\circ = 0.15 \text{ V}$$

$$Fe^{3+}(aq) + e^- \rightarrow Fe^{2+}(aq) \qquad\qquad E^\circ = 0.77 \text{ V}$$

The stronger oxidizing agent is the one involved in the half-reaction with the more positive standard electrode potential, so Fe^{3+} is the stronger oxidizing agent. The oxidation of tin(II) by iron(III) is a spontaneous reaction.

19.71 a. Note that $E^\circ = 0.010$ V and n equals two. Substitute into the equation relating E° and K.

$$E^\circ = \frac{0.0592}{n} \log K$$

$$0.010 \text{ V} = \frac{0.0592}{2} \log K$$

Solving for K, you get

$$\log K = 0.3\underline{3}78$$

Take the antilog of both sides:

$$K = K_c = \text{antilog} (0.3378) = 2.\underline{1}76 = 2.2$$

b. Substitute into the equilibrium expression using $[Sn^{2+}] = x$, and $[Pb^{2+}] = 1.0 \text{ } M - x$.

$$K_c = \frac{[Sn^{2+}]}{[Pb^{2+}]} = \frac{x}{1.0 - x} = 2.176$$

$$x = 0.6\underline{8}51 \text{ } M$$

$$[Pb^{2+}] = 1.0 \text{ } M - x = 0.\underline{3}149 = 0.3 \text{ } M$$

19.73 a. The number of faradays is

$$2.5 \text{ mol Na}^+ \times \frac{1 \text{ mol e}^-}{1 \text{ mol Na}^+} \times \frac{1 \, F}{1 \text{ mol e}^-} = 2.5 \, F$$

The number of coulombs is

$$2.5 \, F \times \frac{9.65 \times 10^4 \, C}{1 \, F} = 2.\underline{4}1 \times 10^5 = 2.4 \times 10^5 \, C$$

b. The number of faradays is

$$2.5 \text{ mol Cu}^{2+} \times \frac{2 \text{ mol e}^-}{1 \text{ mol Cu}^{2+}} \times \frac{1 \, F}{1 \text{ mol e}^-} = 5.0 \, F$$

The number of coulombs is

$$5.0 \, F \times \frac{9.65 \times 10^4 \, C}{1 \, F} = 4.\underline{8}25 \times 10^5 = 4.8 \times 10^5 \, C$$

c. The number of faradays is

$$2.5 \text{ g H}_2\text{O} \times \frac{1 \text{ mol H}_2\text{O}}{18.01 \text{ g H}_2\text{O}} \times \frac{2 \text{ mol e}^-}{1 \text{ mol H}_2\text{O}} \times \frac{1 \, F}{1 \text{ mol e}^-} = 0.2\underline{7}76 = 0.28 \, F$$

The number of coulombs is

$$0.2776 \, F \times \frac{9.65 \times 10^4 \, C}{1 \, F} = 2.\underline{6}79 \times 10^4 = 2.7 \times 10^4 \, C$$

19.75 a. The spontaneous chemical reaction and maximum cell potential are

$$
\begin{array}{lll}
Cd(s) \rightarrow Cd^{2+}(aq) + 2e^- & -E^\circ = 0.40 \text{ V} \\
2Ag^+(aq) + 2e^- \rightarrow 2Ag(s) & E^\circ = 0.80 \text{ V} \\
\hline
Cd(s) + 2Ag^+(aq) \rightarrow 2Ag(s) + Cd^{2+}(aq) & E^\circ_{cell} = 1.20 \text{ V}
\end{array}
$$

b. Addition of S^{2-} would greatly decrease the $Cd^{2+}(aq)$ concentration and help shift the equilibrium to the right, thus forming more Ag.

c. No effect. The size of the electrode makes no difference in the potential.

19.77   Anode: $2H_2O(l) \rightarrow O_2(g) + 4H^+(aq) + 4e^-$

Cathode: $Cu^{2+}(aq) + 2e^- \rightarrow Cu(s)$

19.79 a. First find the moles of silver.

$$2.48 \text{ g Ag} \times \frac{1 \text{ mol Ag}}{107.9 \text{ g Ag}} = 0.02298 \text{ mol Ag}$$

The number of coulombs is

$$0.02298 \text{ mol Ag} \times \frac{1 \text{ mol e}^-}{1 \text{ mol Ag}} \times \frac{9.65 \times 10^4 \text{ C}}{1 \text{ mol e}^-} = 2218 \text{ C}$$

Since amp x sec = coul, the number of seconds is

$$\text{time} = \frac{2218 \text{ C}}{1.50 \text{ amp}} = 1.478 \times 10^3 = 1.48 \times 10^3 \text{ s}$$

b. For the same amount of current, the moles of Cr would be

$$2217 \text{ C} \times \frac{1 \text{ mol e}^-}{9.65 \times 10^4 \text{ C}} \times \frac{1 \text{ mol Cr}}{3 \text{ mol e}^-} \times \frac{52.00 \text{ g Cr}}{1 \text{ mol Cr}}$$

$$= 0.3984 = 0.398 \text{ g}$$

19.81 a. The half-cell reactions, the corresponding half-cell potentials, and their sums are displayed below:

$$Zn(s) \rightarrow Zn^{2+}(aq) + 2e^- \qquad\qquad -E° = 0.76 \text{ V}$$
$$\underline{Cu^{2+}(aq) + 2e^- \rightarrow Cu(s) \qquad\qquad E° = 0.34 \text{ V}}$$
$$Cu^{2+}(aq) + Zn(s) \rightarrow Zn^{2+}(aq) + Cu(s) \qquad\qquad E° = 1.10 \text{ V}$$

Use the Nernst equation to calculate the voltage of the cell.

$$E = E° - \frac{0.0592}{n} \log Q = E° - \frac{0.0592}{n} \log \frac{[Zn^{2+}]}{[Cu^{2+}]}$$

(continued)

Note that n equals two, $[Zn^{2+}] = 0.200\ M$, and $[Cu^{2+}] = 0.0100\ M$.

$$E^\circ = 1.10\ V - \frac{0.0592}{2}\log\frac{[0.200]}{[0.0100]}$$

$$= 1.10\ V - 0.03851\ V = 1.061 = 1.06\ V$$

b. First calculate the moles of electrons passing through the cell.

$$1.00\ \text{amp} \times 225\ \text{s} \times \frac{1\ \text{mol e}^-}{9.65 \times 10^4\ \text{C}} = 0.002332\ \text{mol e}^-$$

The moles of Cu deposited are

$$0.002332\ \text{mol e}^- \times \frac{1\ \text{mol Cu}}{2\ \text{mol e}^-} = 0.001166\ \text{mol Cu}$$

The moles of Cu remaining in the 1.00 L of solution are

$$0.0100 - 0.001166 = 0.008834 = 0.0088\ \text{mol}$$

Since the volume of the solution is 1.00 L, the molarity of Cu^{2+} is 0.0088 M.

19.83 a. Write the cell reaction with the ΔG°_f's beneath

$$Cd(s) + Co^{2+}(aq) \rightarrow Cd^{2+}(aq) + Co(s)$$

| ΔG°_f: | 0 | -54.4 | -77.6 | 0 kJ |

Hence,

$$\Delta G^\circ = \Sigma n\Delta G_f^\circ(\text{products}) - \Sigma m\Delta G_f^\circ(\text{reactants})$$

$$= [-77.6 + 54.4]\ kJ = -23.2\ kJ = -2.32 \times 10^4\ J$$

b. Next determine the standard emf for the cell. Note that n equals two.

$$\Delta G^\circ = -nFE^\circ$$

$$-2.32 \times 10^4\ J = -2(9.65 \times 10^4\ C) \times E^\circ$$

(continued)

Solve for $E°$ to get

$$E° = \frac{-2.32 \times 10^4 \text{ J}}{-2(9.65 \times 10^4 \text{ C})} = 0.12\underline{0}2 = 0.120 \text{ V}$$

The half-reactions and voltages are

$$
\begin{array}{lll}
Cd(s) & \rightarrow & Cd^{2+}(aq) + 2e^- & -E°_{ox} = 0.40 \text{ V}\\
Co^{2+}(aq) + 2e^- & \rightarrow & Co(s) & E°_{red} = ?\\
\hline
Cd(s) + Co^{2+}(aq) & \rightarrow & Cd^{2+}(aq) + Co(s) & E° = 0.1202 \text{ V}
\end{array}
$$

The cell emf is

$$E° = E°_{red} + (-E°_{ox})$$

$$0.1202 \text{ V} = E°_{red} + 0.40 \text{ V}$$

$$E°_{red} = 0.1202 \text{ V} - 0.40 \text{ V} = -0.2\underline{7}9 = -0.28 \text{ V}$$

■ Solutions to Cumulative-Skills Problems

19.85 The half-cell reactions, the corresponding half-cell potentials, and their sums are displayed below:

$$
\begin{array}{lll}
3Zn(s) & \rightarrow & 3Zn^{2+}(aq) + 6e^- & -E° = 0.76 \text{ V}\\
2Cr^{3+}(aq) + 6e^- & \rightarrow & 2Cr(s) & E° = -0.74 \text{ V}\\
\hline
2Cr^{3+}(aq) + 3Zn(s) & \rightarrow & 3Zn^{2+}(aq) + 2Cr(s) & E° = 0.02 \text{ V}
\end{array}
$$

Note that n equals six. Therefore,

$$\Delta G° = -nFE° = -6(9.65 \times 10^4 \text{ C})(0.02 \text{ V}) = -\underline{1}.158 \times 10^4 \text{ J} = -\underline{1}1.58 \text{ kJ}$$

Write the cell reaction with the $\Delta H°_f$'s beneath

$$2Cr^{3+}(aq) + 3Zn(s) \rightarrow 3Zn^{2+}(aq) + 2Cr(s)$$

$\Delta H°_f$:   2(-1971)   0   3(-152.4)   0 kJ

(continued)

Hence,

$$\Delta H° = \Sigma n \Delta H_f°(\text{products}) - \Sigma m \Delta H_f°(\text{reactants})$$

$$= [3(-152.4) - 2(-1971)] \text{ kJ} = 3484.8 = 3485 \text{ kJ}$$

Now calculate $\Delta S°$.

$$\Delta G° = \Delta H° - T\Delta S°$$

$$-11.58 \text{ kJ} = 3484.8 \text{ kJ} - 298 \text{ K} \times \Delta S°$$

Solving for $\Delta S°$ gives

$$\Delta S° = \frac{3484.8 \text{ kJ} + 11.58 \text{ kJ}}{298 \text{ K}} = 11.\underline{7}3 = 11.7 \text{ kJ/K}$$

19.87   The half-cell reactions, the corresponding half-cell potentials, and their sums are displayed below:

$4Br^-(aq) \rightarrow 2Br_2(l) + 4e^-$	$-E° = -1.07$ V	
$O_2(g) + 4H^+(aq) + 4e^- \rightarrow 2H_2O(l)$	$E° = 1.23$ V	
$O_2(g) + 4H^+(aq) + 4Br^-(aq) \rightarrow 2H_2O(l) + 2Br_2(l)$	$E° = 0.16$ V	

Now, convert the pH to $[H^+]$

$$[H^+] = \text{antilog} (-3.60) = 2.\underline{5}1 \times 10^{-4} \text{ } M$$

Under standard conditions $[Br^-] = 1$ M, and the pressure of O_2 is one atm. Thus substitute into the Nernst equation, where n equals four.

$$E = E° - \frac{0.0592}{n} \log Q = E° - \frac{0.0592}{n} \log \frac{1}{[H^+]^4 [Br^-]^4 \times P_{O_2}}$$

$$= 0.16 \text{ V} - \frac{0.0592}{4} \log \frac{1}{(2.51 \times 10^{-4})^4}$$

$$= 0.16 \text{ V} - 0.21\underline{3}1 \text{ V} = -0.0\underline{5}31 = -0.05 \text{ V}$$

Thus the reaction is nonspontaneous at this $[H^+]$.

19.89   Use the K_a to calculate [H+] for the buffer.

$$K_a = \frac{[H^+][OCN^-]}{[HOCN]} = 3.5 \times 10^{-4}$$

Rearrange, and solve for $[H^+]$, assuming [HOCN] and [OCN⁻] remain constant in the buffer. Thus, $[H^+] = K_a = 3.5 \times 10^{-4}\ M$.

In Problem 19.87, $E°$ for this cell reaction was found to be 0.16 V. Under standard conditions [Br⁻] = 1 M, and the pressure of O_2 is one atm. Thus substitute into the Nernst equation, where n equals four.

$$E = E° - \frac{0.0592}{n}\log Q = E° - \frac{0.0592}{n}\log \frac{1}{[H^+]^4[Br^-]^4 \times P_{O_2}}$$

$$= 0.16\ V - \frac{0.0592}{4}\log \frac{1}{(3.5 \times 10^{-4})^4}$$

$$= 0.16\ V - 0.20459\ V = -0.04455 = -0.04\ V$$

Thus the reaction is nonspontaneous at this $[H^+]$.

20. NUCLEAR CHEMISTRY

■ Answers to Review Questions

20.1   The two types of nuclear reactions and their equations are

Radioactive decay: $^{238}_{92}U \rightarrow {}^{234}_{90}Th + {}^{4}_{2}He$

Nuclear bombardment reactions: $^{27}_{13}Al + {}^{4}_{2}He \rightarrow {}^{30}_{15}P + {}^{1}_{0}n$

20.2   To predict whether a nucleus will be stable, look for nuclei that have one of the magic numbers of protons and neutrons. Also look for nuclei that have an even number of protons and an even number of neutrons. Nuclei that fall in the band of stability are also very stable. There are no stable nuclei above atomic number 83.

20.3   The six common types of radioactive decay and the usual condition that leads to each type are listed below (see Table 20.2).

Alpha emission:	$Z > 83$.
Beta emission:	N/Z is too large.
Positron emission:	N/Z is too small.
Electron capture:	N/Z is too small.
Gamma emission:	The nucleus is in an excited state.
Spontaneous fission:	A heavy nucleus with mass number greater than 89.

20.4   The isotopes that begin each of the natural radioactive decay series are uranium-238, uranium-235, and thorium-232.

20.5   The equations are as follows:

   a. $_{7}^{14}\text{N} + _{2}^{4}\text{He} \rightarrow _{8}^{17}\text{O} + _{1}^{1}\text{H}$

   b. $_{13}^{27}\text{Al} + _{2}^{4}\text{He} \rightarrow _{15}^{30}\text{P} + _{0}^{1}\text{n}$

20.6   A curie (Ci) equals 3.700×10^{10} nuclear disintegrations per second.

20.7   It will take cesium-137 three times its half-life of 30.2 y, or 90.6 y, to decay to 1/8 its original mass: 1 to 1/2 to 1/4 to 1/8 = 3 half-lives.

20.8   A radioactive tracer is a radioactive isotope added to a chemical, biological, or physical system to study it. For instance, $^{131}\text{I}^{-}$ is used as a tracer in the study of the dissolving of lead(II) iodide and its equilibrium in a saturated solution.

20.9   Isotope dilution is a technique designed to determine the quantity of a substance in a mixture or to determine the total volume of a solution by adding a known amount of an isotope to it. After removing a portion of the mixture, the fraction by which the isotope has been diluted provides a way of determining the quantity of substance or the total volume of solution.

20.10   Neutron activation analysis is an analysis based on the conversion of stable isotopes to radioactive isotopes by bombarding a sample with neutrons. An unstable nucleus results that then emits gamma rays or radioactive particles (such as beta particles). The amount of stable isotope is proportional to the measured emission.

20.11   The reason the deuteron, $_{1}^{2}\text{H}$, has a mass smaller than the sum of the masses of its constituents is that, when nucleons come together to form a nucleus, energy is released. There must, therefore, be an equivalent decrease in mass because mass and energy are equivalent.

20.12   Two light nuclei, such as two C-12 nuclei, will undergo fusion (with the release of energy) as long as the product nuclei are lighter (which is the case with Na-23 and H-1).

■ Solutions to Practice Problems

Note on significant figures: If the final answer to a solution needs to be rounded off, it is given first with one nonsignificant figure, and the last significant figure is underlined. The final answer is then rounded to the correct number of significant figures. In multiple-step problems, intermediate answers are given with at least one nonsignificant figure; however, only the final answer has been rounded off.

20.19 $^{87}_{37}\text{Rb} \rightarrow \, ^{87}_{38}\text{Sr} + \, ^{0}_{-1}\text{e}$

20.21 $^{232}_{90}\text{Th} \rightarrow \, ^{228}_{88}\text{Ra} + \, ^{4}_{2}\text{He}$

20.23   Let X be the product nucleus. The nuclear equation is

$$^{18}_{9}\text{F} \rightarrow \, ^{A}_{Z}\text{X} + \, ^{0}_{1}\text{e}$$

From the superscripts: $18 = A + 0$, or $A = 18$; from the subscripts: $9 = Z + 1$, or $Z = 8$. Thus the product of the reaction is $^{18}_{8}\text{X}$, and because element 8 is oxygen, symbol O, the nuclear equation is

$$^{18}_{9}\text{F} \rightarrow \, ^{18}_{8}\text{O} + \, ^{0}_{1}\text{e}$$

20.25 a.   Neither nucleus has an atomic number that is a magic number of protons. Find how many neutrons are in each nucleus.

     Sb: No. of neutrons $= A - Z = 122 - 51 = 71$

     Xe: No. of neutrons $= A - Z = 136 - 54 = 82$

Because 82 is a magic number for neutrons (implying stability of nucleus), you predict $^{136}_{54}\text{Xe}$ is stable and $^{122}_{51}\text{Sb}$ is radioactive.

   b. $^{204}_{82}\text{Pb}$ has a magic number of protons (82), so it is expected to be the stable nucleus, and $^{204}_{85}\text{At}$ is radioactive (atomic number greater than 83).

(continued)

c. Rb does not have an atomic number that is a magic number of protons. Find the numbers of neutrons in the two isotopes.

$${}^{87}_{37}\text{Rb}:$$ No. of neutrons = 87 - 37 = 50

$${}^{80}_{37}\text{Rb}:$$ No. of neutrons = 80 - 37 = 43

Because 50 is a magic number for neutrons, you predict $${}^{87}_{37}\text{Rb}$$ is stable, and $${}^{80}_{37}\text{Rb}$$ is radioactive.

20.27 a. α-emission is most likely for nuclei with $Z > 83$

b. positron emission (more likely; $Z < 20$) or electron capture because mass no. is less than that of Cu = 63.5

c. β-emission (mass no. > He = 4)

20.29 $$\frac{12.6 \text{ MeV}}{1 \text{ proton}} = \frac{12.6 \times 10^6 \text{ eV}}{1 \text{ proton}} \times \frac{1.602 \times 10^{-19} \text{ J}}{1 \text{ eV}} \times \frac{1 \text{ kJ}}{10^3 \text{ J}} \times \frac{6.02 \times 10^{23} \text{ protons}}{1 \text{ mol}}$$

$$= 1.2\underline{1}5 \times 10^9 = 1.22 \times 10^9 \text{ kJ/mol}$$

20.31 $${}^{6}_{3}\text{Li} + {}^{1}_{0}\text{n} \rightarrow {}^{A}_{Z}\text{X} + {}^{3}_{1}\text{H}$$

From the superscripts: $6 + 1 = A + 3$, or $A = 6 + 1 - 3 = 4$; from the subscripts: $3 + 0 = Z + 1$, or $Z = 3 - 1 = 2$. The product nucleus is $${}^{4}_{2}\text{X}$$. The element with $Z = 2$ is helium (He), so the missing nuclide is $${}^{4}_{2}\text{He}$$.

20.33 The reaction may be written

$${}^{249}_{98}\text{Cf} + {}^{A}_{Z}\text{X} \rightarrow {}^{257}_{104}\text{Rf} + 4{}^{1}_{0}\text{n}$$

From the superscripts: $249 + A = 257 + (4 \times 1)$, or $A = 257 + 4 - 249 = 12$; from the subscripts: $98 + Z = 104 + 0$, or $Z = 104 - 98 = 6$. The element with $Z = 6$ is carbon (the bombarding nuclei were $${}^{12}_{6}\text{C}$$).

20.35 The rate of decay is 8.94×10^{10} nuclei/s. The number of nuclei in the sample is

$$0.250 \times 10^{-3} \text{ g H-3} \times \frac{1 \text{ mol H-3}}{3.02 \text{ g H-3}} \times \frac{6.02 \times 10^{23} \text{ nuclei}}{1 \text{ mol H-3}} = 4.9\underline{8}3 \times 10^{19} \text{ nuclei}$$

The rate equation is rate = kN_t. Solve for k.

$$k = \frac{\text{rate}}{N_t} = \frac{8.94 \times 10^{10} \text{ nuclei/s}}{4.98 \times 10^{19} \text{ nuclei}} = 1.7\underline{9}4 \times 10^{-9} = 1.79 \times 10^{-9} \text{ s}^{-1}$$

20.37 $t_{1/2} = \dfrac{0.693}{k} = \dfrac{0.693}{1.7 \times 10^{-21} \text{ /s}} \times \dfrac{1 \text{ h}}{3600 \text{ s}} \times \dfrac{1 \text{ d}}{24 \text{ h}} \times \dfrac{1 \text{ y}}{365 \text{ d}} = 1.\underline{2}92 \times 10^{13}$

$$= 1.3 \times 10^{13} \text{ y}$$

20.39 Find k from the half-life.

$$k = \frac{0.693}{t_{1/2}} = \frac{0.693}{2.69 \text{ d}} \times \frac{1 \text{ d}}{24 \text{ h}} \times \frac{1 \text{ h}}{3600 \text{ s}} = 2.9\underline{8}2 \times 10^{-6} \text{/s}$$

Before substituting into the rate equation, find the number of gold nuclei from the mass.

$$0.86 \times 10^{-3} \text{ g Au-198} \times \frac{1 \text{ mol Au-198}}{198 \text{ g Au-198}} \times \frac{6.02 \times 10^{23} \text{ Au-198 nuclei}}{1 \text{ mol Au-198}}$$

$$= 2.\underline{6}1 \times 10^{18} \text{ Au-198 nuclei}$$

Now find the rate.

$$\text{Rate} = kN_t = (2.982 \times 10^{-6} \text{/s})(2.61 \times 10^{18} \text{ nuclei}) = 7.\underline{7}9 \times 10^{12} \text{ nuclei/s}$$

$$\text{Activity} = 7.79 \times 10^{12} \text{ nuclei/s} \times \frac{1 \text{ Ci}}{3.700 \times 10^{10} \text{ nuclei /s}} = 2\underline{1}0.7 = 2.1 \times 10^{2} \text{ Ci}$$

20.41   Find k from the half-life.

$$k = \frac{0.693}{t_{1/2}} = \frac{0.693}{14.3\ d} \times \frac{1\ d}{24\ h} \times \frac{1\ h}{3600\ s} = 5.6\underline{0}9 \times 10^{-7}/s$$

Solve the rate equation for N_t.

$$N_t = \frac{rate}{k} = \frac{6.0 \times 10^{12}\ nuclei/s}{5.609 \times 10^{-7}\ /s} = 1.\underline{0}7 \times 10^{19}\ nuclei$$

Convert N_t to the mass of P-32.

$$1.07 \times 10^{19}\ \text{P-32 nuclei} \times \frac{1\ mol\ \text{P-32}}{6.02 \times 10^{23}\ \text{P-32 nuclei}} \times \frac{32\ g\ \text{P-32}}{1\ mol\ \text{P-32}}$$

$$= 5.\underline{6}8 \times 10^{-4} = 5.7 \times 10^{-4}\ g\ \text{P-32}$$

20.43   Substituting for $k = 0.693/t_{1/2}$ gives

$$\ln\frac{N_t}{N_0} = -kt = \frac{-0.693\ t}{t_{1/2}} = \frac{-0.693\ (12.0\ h)}{15.0\ h} = -0.55\underline{4}4$$

Taking the antilog of both sides of this equation gives

$$\frac{N_t}{N_0} = e^{-0.5544} = 0.57\underline{4}41$$

After 12.0 h, 57.4 percent of the Na-24 remains.

$$5.0\ \mu g \times 0.57441 = 2.\underline{8}72\ \mu g = 2.9\ \mu g$$

20.45   After 1.97 sec, the amount of nitrogen-17 remaining is 100% - 28.0% = 72.0 percent, so $N_t/N_0 = 0.720$.

$$\ln\frac{N_t}{N_0} = -kt = \frac{-0.693\ t}{t_{1/2}}$$

$$\ln\,(0.720) = \frac{-0.693(1.97\ s)}{t_{1/2}}$$

(continued)

Solve for $t_{1/2}$.

$$t_{1/2} = \frac{-0.693(1.97 \text{ s})}{\ln(0.720)} = 4.1\underline{5}58 = 4.16 \text{ s}$$

20.47 Use the equation for the number of nuclei in a sample after a time, t.

$$\ln\frac{N_t}{N_0} = -kt = \frac{-0.693\,t}{t_{1/2}}$$

Rearrange this to give an expression for t.

$$t = \frac{t_{1/2}}{0.693}\ln\frac{N_0}{N_t}$$

At $t = 0$, the rate is 15.3 nuclei/s, and at a later time t, the rate is 8.1 nuclei/s. Since the half-life for carbon-14 is 5730 years,

$$t = \frac{5730 \text{ y}}{0.693}\ln\left(\frac{15.3}{8.1}\right) = 5.\underline{2}58 \times 10^3 = 5.3 \times 10^3 \text{ y}$$

20.49 Write the atomic masses below each nuclide symbol and calculate Δm.

$$^{2}_{1}\text{H} \quad + \quad ^{3}_{1}\text{H} \quad \rightarrow \quad ^{4}_{2}\text{He} \quad + \quad ^{1}_{0}\text{n}$$

Masses: 2.01400   3.01605   4.00260   1.008665 amu

$\Delta m = [4.00260 + 1.008665] - [2.01400 + 3.01605] = -0.0187\underline{8}5$

The energy change for one mole is

$$\Delta E = (\Delta m)c^2 = (-0.018785 \times 10^{-3} \text{ kg})(2.998 \times 10^8 \text{ m/s})^2$$

$$= -1.68\underline{8}39 \times 10^{12} \text{ kg}\bullet\text{m}^2/\text{s}^2 = -1.688 \times 10^{12} \text{ J}$$

Finally, calculate the energy change in MeV for one $^{2}_{1}\text{H}$ nucleus.

$$\frac{-1.68839 \times 10^{12} \text{ J}}{1 \text{ mol}} \times \frac{1 \text{ mol}}{6.022 \times 10^{23} \text{ nuclei}} \times \frac{1 \text{ MeV}}{1.602 \times 10^{-13} \text{ J}}$$

$$= -17.5\underline{0}1 = -17.50 \text{ MeV}$$

20.51   Mass of three protons = 3 x 1.00728 amu = 3.02184 amu

Mass of three neutrons = 3 x 1.008665 amu = 3.025995 amu

Total mass of nucleons = (3.02184 + 3.025995) amu = 6.047835 amu

Mass defect = total nucleon mass - nuclear mass = (6.047835 - 6.01512) amu

= 0.032715 amu

$$\Delta E = (\Delta m)c^2 = 0.032715 \text{ amu} \times \frac{1 \text{ g}}{6.022 \times 10^{23} \times 1 \text{ amu}} \times \frac{1 \text{ kg}}{10^3 \text{ g}}$$

$$\times (2.998 \times 10^8 \text{ m/s})^2 \times \frac{1 \text{ MeV}}{1.602 \times 10^{-13} \text{ J}} = 30.479 \text{ MeV}$$

$$\text{Binding energy per nucleon} = \frac{30.479 \text{ MeV}}{6 \text{ nucleons}} = 5.0799 = 5.080 \text{ MeV/nucleon}$$

■ Solutions to General Problems

20.53   F-18, having fewer neutrons than the stable F-19, is expected to decay to a nucleus with a lower atomic number (and hence a higher N/Z ratio) by electron capture or positron emission. F-21, having more neutrons than the stable isotope, is expected to decay by beta emission to give a nucleus with a higher atomic number (and hence a lower N/Z ratio).

20.55 $^{209}_{83}\text{Bi} + {}^{4}_{2}\text{He} \rightarrow {}^{A}_{85}\text{At} + 2\,{}^{1}_{0}\text{n}$

From the superscripts: 209 + 4 = A + (2 x 1), or A = 213 - 2 = 211.

The reaction is

$^{209}_{83}\text{Bi} + {}^{4}_{2}\text{He} \rightarrow {}^{211}_{85}\text{At} + 2\,{}^{1}_{0}\text{n}$

20.57 $^{238}_{92}Bi + ^{12}_{6}C \rightarrow ^{A}_{Z}X + 4\,^{1}_{0}n$

From the superscripts: $238 + 12 = A + (4 \times 1)$, or $A = 250 - 4 = 246$; from the subscripts: $92 + 6 = Z + (4 \times 0)$, or $Z = 98$.

The element with z = 98 is californium (Cf), so the equation is

$^{238}_{92}Bi + ^{12}_{6}C \rightarrow ^{246}_{98}Cf + 4\,^{1}_{0}n$

20.59 Use the equation for the number of nuclei in a sample after a time, t.

$$\ln\frac{N_t}{N_0} = \frac{-0.693\,t}{t_{1/2}}$$

Rearrange this to give an expression for t.

$$t = \frac{t_{1/2}}{0.693}\ln\frac{N_0}{N_t}$$

Substituting

$$t = \frac{(12.3\,y)}{0.693}\ln\frac{N_0}{0.70\,N_0} = 6.\underline{3}3 = 6.3\,y$$

20.61 Write the atomic masses below each nuclide symbol and calculate Δm.

$^{1}_{0}n + ^{235}_{92}U \rightarrow ^{136}_{53}I + ^{96}_{39}Y + 4\,^{1}_{0}n$

Masses: 1.008665   235.04392   135.8401   95.8629   1.008665 amu

$\Delta m = [135.8401 + 95.8629 + 4 \times 1.008665 - (1.008665 + 235.04392)]$ amu

$= -0.314\underline{9}25$ amu

When one mol of U-235 decays, the change in mass is -0.314925 g. Therefore, for 5.00 kg of U-235, the change in mass is

5.00×10^3 g U-235 $\times \dfrac{1\text{ mol U-235}}{235.04392\text{ g U-235}} \times \dfrac{-0.314925\text{ g}}{1\text{ mol U-235}} = -6.6\underline{9}927$ g

(continued)

Converting mass to energy gives:

$$E = (\Delta m)c^2 = (-6.69927 \times 10^{-3} \text{ kg})(2.998 \times 10^8 \text{ m/s})^2 = -6.0\underline{2}13 \times 10^{14} \text{ J}$$

For the combustion of carbon:

C(graphite)	+	$O_2(g)$	$\rightarrow$	$CO_2(g)$
ΔH_f° 0		0		-393.5 kJ/mol

For the reaction, $\Delta H = \Delta E = -393.5$ kJ/mol.

$$5.00 \times 10^3 \text{ g} \times \frac{1 \text{ mol C}}{12.01 \text{ g C}} \times \frac{-393.5 \text{ kJ}}{1 \text{ mol C}} = -1.6\underline{3}8 \times 10^5 = -1.64 \times 10^5 \text{ kJ}$$

The energy released in the fission of 5.00 kg of U-235 (6.02×10^{11} kJ) is larger by several orders of magnitude than the energy released by burning 5.00 kg of C (1.64×10^5 kJ).

20.63 a. The balanced equation is

$$^{47}_{20}\text{Ca} \rightarrow \, ^{47}_{21}\text{Sc} + \, ^{0}_{-1}\beta$$

b. Use the rate law to find the initial amount. The time is 48 hours (2.0 d), and the half-life is 4.536 d. Substituting

$$\ln \frac{N_t}{N_0} = -kt = \frac{-0.693 \, t}{t_{1/2}} = \frac{-0.693 \, (2.0 \text{ d})}{4.536 \text{ d}} = -0.3\underline{0}55$$

Taking the antilog of each side gives

$$\frac{N_t}{N_0} = \frac{A_t}{A_0} = e^{-0.3055} = 0.7\underline{3}67$$

Since A_t is 25.0 μg, you can solve for A_0.

$$A_0 = \frac{25.0 \, \mu\text{g}}{0.7367} = 33.\underline{9}3 = 34 \, \mu\text{g Ca-47}$$

Convert this to mass of $CaSO_4$.

$$33.93 \, \mu\text{g Ca-47} \times \frac{136.15 \text{ g CaSO}_4}{40.08 \text{ g Ca}} = 11\underline{5}.2 = 1.2 \times 10^2 \, \mu\text{g}$$

■ Solutions to Cumulative-Skills Problems

20.65 Calculate the number of P-32 nuclei in the sample (N_t). For this, you need the formula weight of N_3PO_4 containing 15.6 percent P-32 and $(100.0 - 15.6)\% = 84.4$ percent naturally occurring P. The formula weight of Na_3PO_4 containing naturally occurring P is 163.9 amu; the formula weight of Na_3PO_4 with 100 percent P-32 is 165.0. The formula weight of Na_3PO_4 containing 15.6 percent P-32 is obtained from the weighted average of the formula weights:

$$(163.9 \text{ amu} \times 0.844) + (165.0 \text{ amu} \times 0.156) = 164.1 \text{ amu}$$

The moles of P in the sample equal

$$0.0545 \text{ g Na}_3PO_4 \times \frac{1 \text{ mol Na}_3PO_4}{164.1 \text{ g Na}_3PO_4} \times \frac{1 \text{ mol P}}{1 \text{ mol Na}_3PO_4} = 3.321 \times 10^{-4} \text{ mol P}$$

and the moles of P-32 equal

$$(3.321 \times 10^{-4} \text{ mol P}) \times 0.156 = 5.181 \times 10^{-5} \text{ mol P-32}$$

Then the number of P-32 nuclei is

$$(5.181 \times 10^{-5} \text{ mol P-32}) \times \frac{6.022 \times 10^{23} \text{ P-32 nuclei}}{1 \text{ mol P-32}} = 3.120 \times 10^{19} \text{ P-32 nuclei}$$

Now calculate the rate of disintegrations. Because the rate = kN_t, first find the value of k in reciprocal seconds:

$$k = \frac{0.693}{t_{1/2}} = \frac{0.693}{14.3 \text{ d}} \times \frac{1 \text{ d}}{24 \text{ h}} \times \frac{1 \text{ h}}{3600 \text{ s}} = 5.609 \times 10^{-7} \text{ /s}$$

Therefore, the rate of disintegrations is

$$\text{Rate} = kN_t = (5.609 \times 10^{-7}\text{/s}) \times (3.120 \times 10^{19} \text{ P-32 nuclei}) = 1.750 \times 10^{13}$$

$$= 1.75 \times 10^{13} \text{ P-32 nuclei/s}$$

20.67 $2p + 2n \rightarrow$ He-4

On a mole basis, the mass difference, Δm, is

$\Delta m = [4.00260 - 2(1.00728) - 2(1.00867)]$ g/mol $= -0.02930$ g/mol

$\qquad = -2.930 \times 10^{-5}$ kg/mol

The energy evolved per mole is

$\Delta E = (\Delta m)c^2 = -2.930 \times 10^{-5}$ kg/mol x $(2.998 \times 10^8$ m/s$)^2$

$\qquad = -2.63\underline{2}58 \times 10^{12}$ kg·m^2/s$^2 = -2.63\underline{2}58 \times 10^{12}$ J $= -2.63\underline{2}58 \times 10^9$ kJ/mol

Next calculate $\Delta H°$ for burning of ethane:

$\qquad$ C$_2$H$_6$(g)   +   7/2O$_2$ $\rightarrow$ 2CO$_2$(g)   +   3H$_2$O(g)

$\Delta H_f°$ = -84.68 $\qquad\qquad$ 0 $\qquad\qquad$ 2(-393.5) $\qquad$ 3(-241.8) kJ

$\Delta H° = [2(-393.\underline{5}) + 3(-241.8) - (-84.68)]$ kJ $= -1427.\underline{7}2$ kJ/mol ethane

Now calculate the mol of ethane needed to obtain 2.63258×10^9 kJ heat:

$\qquad -2.63\underline{2}58 \times 10^9$ kJ x $\dfrac{1 \text{ mol ethane}}{-1427.72 \text{ kJ}}$ $= 1.843\underline{9}0 \times 10^6$ mol ethane

Finally, convert moles to liters at 25°C and 725 mmHg.

$$V = \frac{nRT}{P} = \frac{(1.84390 \times 10^6 \text{ mol})\,[0.0821 \text{ L} \cdot \text{atm/(K} \cdot \text{mol)}]\,(298 \text{ K})}{(725/760) \text{ atm}}$$

$$= 4.7\underline{2}9 \times 10^7 = 4.73 \times 10^7 \text{ L}$$

21. CHEMISTRY OF THE METALS

■ Answers to Review Questions

21.1 An alloy is a material with metallic properties that is either a compound or a mixture. If the alloy is a mixture, it may be homogeneous (a solution) or heterogeneous. Gold jewelry is made from an alloy that is a solid solution of gold containing some silver.

21.2 A metal is a material that is lustrous (shiny), has high electrical and heat conductivities, and is malleable and ductile.

21.3 The usual compounds from which metals are obtained are oxides (and hydroxides), sulfides, and carbonates.

21.4 The basic steps in the production of a pure metal from a natural source are:
(1) Preliminary treatment: separating the metal-containing mineral from the less desirable parts of the ore. The mineral may also be transformed by chemical reaction to a metal compound that is more easily reduced to the free metal; (2) Reduction: The metal compound is reduced to the free metal by electrolysis or chemical reduction; (3) Refining: The free metal is purified.

(1) Preliminary treatment: Aluminum oxide is obtained from bauxite by the Bayer process. Bauxite contains aluminum hydroxide, aluminum oxide hydroxide, and other worthless constituents. It is mixed with hot, aqueous sodium hydroxide solution, which dissolves the amphoteric aluminum minerals along with some silicates. When the solution is cooled, $Al(OH)_3$ precipitates, leaving the silicates behind. The aluminum hydroxide is finally calcined (heated strongly in a furnace) to produce purified aluminum oxide, Al_2O_3. (2) Reduction: Aluminum is then obtained by reduction using the Hall-Héroult process, which is the electrolysis of a molten mixture of aluminum oxide in cryolyte (Na_3AlF_6). (3) Refining: The aluminum can be further refined and purified.

21.5 Tungsten reacts with carbon to produce tungsten carbide, WC, which is unreactive. Instead, the metal is prepared from tungsten (VI) oxide obtained from the processing of tungsten ore. The oxide is reduced by heating it in a stream of hydrogen gas.

21.6   In a metal, the outer orbits of an enormous number of metal atoms overlap to form an enormous number of molecular orbitals that are delocalized over the metal. As a result, a large number of energy levels are crowded together into bands. These bands are half-filled with electrons. When a voltage is applied to the metal crystal, electrons are excited to the unoccupied orbitals and move toward the positive pole of the voltage source. This is electrical conductivity.

21.7   The reactions are

$$2Li(s) + 2H_2O(l) \rightarrow 2LiOH(aq) + H_2(g)$$

$$4Li(s) + O_2(g) \rightarrow 2Li_2O(s)$$

$$2Na(s) + 2H_2O(l) \rightarrow 2NaOH(aq) + H_2(g)$$

$$2Na(s) + O_2(g) \rightarrow Na_2O_2(s)$$

21.8   The reaction is $2Na(s) + 2C_2H_5OH(l) \rightarrow H_2(g) + 2NaOC_2H_5(aq)$.

21.9   Sodium hydroxide is manufactured by the electrolysis of aqueous sodium chloride, which also produces chlorine gas as a major product.

21.10   The uses are given after each compound: sodium chloride–used for making sodium hydroxide and in seasoning; sodium hydroxide–used in aluminum production and in producing sodium compounds such as soap; and sodium carbonate–used to make glass and as washing soda in many detergent preparations.

21.11 a. Calcium oxide is prepared industrially from calcium carbonate:

$$CaCO_3(s) \rightarrow CaO(s) + CO_2(g)$$

b. Calcium hydroxide is prepared from the reaction of calcium oxide and water:

$$CaO(s) + H_2O(l) \rightarrow Ca(OH)_2(aq)$$

21.12 $CaCO_3(s) + 2HCl(aq) \rightarrow CO_2(g) + CaCl_2(aq) + H_2O(l)$

21.13   Some major uses of aluminum oxide are making abrasives for grinding tools, fusing with small amounts of other metal oxides to make synthetic sapphires and rubies, and making industrial ceramics.

21.14   To purify municipal water, aluminum sulfate and calcium hydroxide are added to waste water, forming a gelatinous precipitate of aluminum hydroxide. Colloidal particles of clay (usually present in the waste water) and other substances adhere to the aluminum hydroxide, whose particles are large enough to be filtered from the water to purify it.

21.15 Characteristics of the transition elements that set them apart from the main-group elements are the following: (1) The transition elements are metals with high melting points (only the IIB elements have low melting points). Most main-group elements have low melting points. (2) Each of the transition metals has several oxidation states (except for the IIIB and IIB elements). Most main-group metals have only one oxidation state in addition to zero. (3) Transition-metal compounds are often colored, and many are paramagnetic. Most main-group compounds are colorless and diamagnetic.

21.16 Technetium has the electron configuration [Kr] $4d^5 5s^2$.

21.17 Werner showed that the electrical conductance of a solution of $[Pt(NH_3)_4Cl_2]Cl_2$ corresponded to that of three ions in solution and that two of the chloride ions could be precipitated as AgCl whereas the other two could not.

21.18 A complex ion is a metal atom or ion with Lewis bases attached to it through coordinate covalent bonds. A ligand is a Lewis base attached to a metal ion in a complex; it may be either a molecule or an anion, rarely a cation. The coordination number of a metal atom in a complex ion is the total number of bonds the metal forms with ligands. An example of a complex ion is $Fe(CN)_6^{4-}$; an example of a ligand is CN^-; and the coordination number of the preceding complex ion is six.

21.19 A bidentate ligand is a ligand that bonds to a metal ion through two atoms. Two examples are ethylenediamine, $H_2N-C_2H_4-NH_2$, and the oxalate ion, $^-O_2C-CO_2^-$.

21.20

21.21 The three properties are isomerism, paramagnetism, and color (or absorption of visible and ultraviolet radiation).

21.22 According to valence bond theory, a ligand orbital containing two electrons overlaps an unoccupied orbital on the metal ion.

21.23 a. A high-spin Fe(II) octahedral complex is

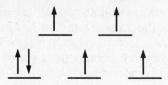

b. A low-spin Fe(II) octahedral complex is

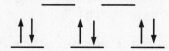

21.24   Crystal field splitting is the difference in energy between the two sets of d orbitals for a given structure (such as octahedral) in complex ions. It is determined experimentally by measuring the energy of light absorbed by complex ions.

21.25   The spectrochemical series is the arrangement of ligands in order of the relative size of the crystal field splittings (Δ) they induce in the d orbitals of a given oxidation state of a given metal ion. The order is the same no matter what metal or oxidation state is involved. For CN⁻, H_2O, Cl⁻, and NH_3, the order of increasing crystal field splitting is

Cl⁻ < H_2O < NH_3 < CN⁻

where CN⁻ always acts as a strong-bonding ligand.

21.26   Pairing energy, P, is the energy required to place two electrons in the same orbital. If the crystal field splitting (Δ) is small because of weak-bonding ligands, then the pairing energy will be larger, and the complex will be high-spin. If the crystal field splitting (Δ) is large because of strong-bonding ligands, then the pairing energy will be smaller, and the complex will be low-spin.

■ Solutions to Practice Problems

Note on significant figures: If the final answer to a solution needs to be rounded off, it is given first with one nonsignificant figure, and the last significant figure is underlined. The final answer is then rounded to the correct number of significant figures. In multiple-step problems, intermediate answers are given with at least one nonsignificant figure; however, only the final answer has been rounded off.

21.35 In the reaction, three mol of H_2 are used to form two mol of iron. Using the respective atomic masses of 55.85 g/mol for Fe and 2.016 g/mol for H_2, you can calculate the mass of Fe as follows:

$$2.00 \times 10^3 \text{ g } H_2 \times \frac{1 \text{ mol } H_2}{2.016 \text{ g } H_2} \times \frac{2 \text{ mol Fe}}{3 \text{ mol } H_2} \times \frac{55.85 \text{ g Fe}}{1 \text{ mol Fe}}$$

$$= 3.6\underline{9}3 \times 10^4 = 3.69 \times 10^4 \text{ g} = 36.9 \text{ kg Fe}$$

21.37 The equation with $\Delta H°_f$'s recorded beneath each substance is

$$PbS(s) \quad + \quad 3/2 O_2(g) \quad \rightarrow \quad PbO(s) \quad + \quad SO_2(g)$$

$$\text{-98.32} \qquad\qquad 0 \qquad\qquad\quad \text{-219.4} \qquad\quad \text{-296.8 (kJ)}$$

$$\Delta H° = [(-219.4) + (-296.8) -(-98.32)] \text{ kJ} = -417.\underline{8}8 = -417.9 \text{kJ (exothermic)}$$

21.39 a. $2K(s) + Br_2(l) \rightarrow 2KBr(s)$

b. $2K(s) + 2H_2O(l) \rightarrow 2KOH(aq) + H_2(g)$

c. $Li_2CO_3(aq) + 2HNO_3(aq) \rightarrow H_2O(l) + 2LiNO_3(aq) + CO_2(g)$

d. $K_2SO_4(aq) + Pb(NO_3)_2(aq) \rightarrow PbSO_4(s) + 2KNO_3(aq)$

21.41 $Ca(OH)_2$ can be identified directly by adding an anion that will precipitate the Ca^{2+} (and not the Na^+). For example, adding CO_3^{2-} will precipitate $CaCO_3$ but will not precipitate Na^+ ion.

21.43 a. $Ba(s) + 2H_2O(l) \rightarrow Ba(OH)_2(aq) + H_2(g)$

b. $Mg(OH)_2(s) + 2HNO_3(aq) \rightarrow 2H_2O(l) + Mg(NO_3)_2(aq)$

c. $Mg(s) + NiCl_2(aq) \rightarrow Ni(s) + MgCl_2(aq)$

d. $2NaOH(aq) + MgSO_4(aq) \rightarrow Mg(OH)_2(s) + Na_2SO_4(aq)$

21.45   The half-reactions and their sum are as follows:

$$PbO_2 + 4H^+ + 2e^- \rightarrow Pb^{2+} + 2H_2O$$

$$2Cl^- \rightarrow Cl_2 + 2e^-$$

$$PbO_2 + 4H^+ + 2Cl^- \rightarrow Pb^{2+} + Cl_2 + 2H_2O$$

Adding $2Cl^-$ to both sides gives

$$PbO_2 + 4HCl \rightarrow PbCl_2 + Cl_2 + 2H_2O$$

21.47 a. $Al_2O_3(s) + 3H_2SO_4(aq) \rightarrow Al_2(SO_4)_3(aq) + 3H_2O(l)$

b. $Al(s) + 3AgNO_3(aq) \rightarrow 3Ag(s) + Al(NO_3)_3(aq)$

c. $Pb(NO_3)_2(aq) + 2NaI(aq) \rightarrow PbI_2(s) + 2NaNO_3(aq)$

d. $8Al(s) + 3Mn_3O_4(s) \rightarrow 9Mn(s) + 4Al_2O_3(s)$

21.49 a.   The charge on the carbonate ion is -2. For $FeCO_3$ to be neutral, the oxidation number of iron must be +2.

b.   The oxidation number of oxygen is -2. For the sum of the oxidation numbers of all atoms to be zero, manganese must be in the +4 oxidation state.

c.   The oxidation number of the chlorine is -1, so the oxidation number of copper must be +1.

d.   The oxidation number of oxygen is -2. The oxidation number of chlorine is -1. For the sum of the oxidation numbers to be zero, the oxidation number of chromium must be +6.

$$+6 + 2(-2) + 2(-1) = 0$$

21.51 a.   Four. There are four cyanide groups coordinated to the gold atom.

b.   Six. There are four ammonia molecules and two water molecules coordinated to the cobalt.

c.   Four. Each of the ethylenediamine molecules bonds to the gold atom through two nitrogen atoms.

d.   Six. Each ethylenediamine molecule bonds to the chromium atom through two nitrogen atoms, and the oxalate ion bonds to the chromium through two oxygen atoms.

21.53 a. The charge on the $Ni(CN)_4^{2-}$ ion is -2 to balance the charge of +2 from the two K^+ ions. Each cyanide ion has a charge of -1. The sum of the oxidation number of nickel and the charge on the cyanide ions must equal the charge, so

$$-2 = [4 \times (-1)] + 1 \times [\text{ox. no. (Ni)}]$$

$$\text{Ox. no. (Ni)} = -2 - (-4) = +2$$

b. The charge on ethylenediamine is zero, so the oxidation number of Mo is equal to the charge on the complex ion.

$$\text{Ox. no. (Mo)} = +3$$

c. Oxalate ion has a charge of -2, so

$$\text{Ox. no. of Cr} = \text{charge of complex ion} \ -3 \times (\text{charge of oxalate ion})$$
$$= (-3) - [3 \times (-2)] = +3$$

d. Chloride ion has a charge of -1, so the charge on the complex ion is +2. The NH_3 ligands are neutral and contribute nothing to the charge of the complex ion.

$$\text{Ox. no. of Co} = \text{charge of complex ion} \ - \ \text{charge of nitrite ligand}$$
$$= +2 - (-1) = +3$$

21.55 a.

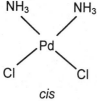

cis

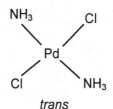

trans

b. No geometric isomerism.

c. No geometric isomerism.

(continued)

d.

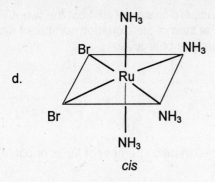

cis

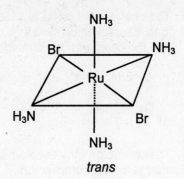

trans

21.57 a. V^{3+} has two *d* electrons arranged as shown:

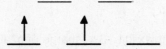

There are two unpaired electrons.

b. Co^{2+} has seven *d* electrons. In the high-spin case, they are arranged as follows:

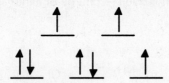

There are three unpaired e⁻.

c. Mn^{3+} has four *d* electrons. In the low-spin case, they are arranged as follows:

There are two unpaired e⁻.

21.59 Purple

■ Solutions to General Problems

21.61 The equations are

$$2HCl + Mg(OH)_2 \rightarrow MgCl_2 + 2H_2O$$

$$HCl + NaOH \rightarrow NaCl + H_2O.$$

Using a 1:1 ratio of HCl to NaOH, calculate the mol of HCl reacting with NaOH:

$$\left[\frac{0.4987 \text{ mol HCl}}{L} \times 0.05000 \text{ L} \right] - \left[\frac{0.2456 \text{ mol NaOH}}{L} \times 0.03942 \text{ L} \right]$$

$$= 0.015253 \text{ mol HCl}$$

The mass of $Mg(OH)_2$ is

$$0.015253 \text{ mol HCl} \times \frac{1 \text{ mol Mg(OH)}_2}{2 \text{ mol HCl}} \times \frac{58.33 \text{ g Mg(OH)}_2}{1 \text{ mol Mg(OH)}_2}$$

$$= 0.44486 \text{ g Mg(OH)}_2$$

The mass percentage is

$$\frac{0.44486 \text{ g}}{5.436 \text{ g}} \times 100\% = 8.1837 = 8.184 \text{ percent}$$

21.63 The color in transition-metal complexes is due to absorption of light when a *d* electron moves to a higher energy level. Because Sc^{3+} has no *d* electrons, it is expected to be colorless.

22. CHEMISTRY OF THE NONMETALS

■ Answers to Review Questions

22.1   In the steam-reforming process, steam and hydrocarbons from natural gas or petroleum react at high temperature and pressure in the presence of a catalyst to form carbon monoxide and hydrogen. A typical reaction involving propane (C_3H_8) is

$$C_3H_8(g) + 3H_2O(g) \xrightarrow[\Delta]{Ni} 3CO(g) + 7H_2(g)$$

The carbon monoxide is removed from the mixture by reacting with steam in the presence of a catalyst to give carbon dioxide and more hydrogen.

$$CO(g) + H_2O(g) \xrightarrow[\Delta]{Catalyst} CO_2(g) + H_2(g)$$

Finally, the carbon dioxide is removed by dissolving it in a basic aqueous solution.

22.2   The combustion of hydrogen produces more heat per gram than the combustion of any other fuel (120 kJ/g). Unlike hydrocarbons, it is a clean fuel since the product is environmentally benign water. These features, in the face of a dwindling supply of hydrocarbons, indicate hydrogen gas may become the favorite fuel of the twenty-first century.

22.3   A binary hydride is a compound that contains hydrogen and one other element. There are three categories of binary hydrides. The first type is an ionic hydride that contains the hydride ion, H⁻, and is formed via reaction with an alkali metal or larger group IIA metal. An example is LiH. A second type is a covalent hydride that is a molecular compound in which hydrogen is covalently bonded to another element. An example is NH_3. The third type is a metallic hydride formed from a transition metal and hydrogen. These compounds contain hydrogen spread throughout a metal crystal and occupying the holes in the crystal lattice, sometimes in nonstoichiometric amounts. Thus the composition is variable. An example is $TiH_{1.7}$.

22.4   Three allotropes of carbon are diamond, graphite, and fullerenes. Diamond is a covalent network solid in which each carbon is tetrahedrally (sp^3) bonded to four other carbon atoms. Graphite has a layer structure with each layer being attracted to one another by London forces. Within a layer, each carbon atom is covalently bonded to three other atoms, giving a flat layer of carbon-atom hexagons. The bonding is sp^2 with delocalized π bonds. Fullerenes are molecules with a closed cage of carbon atoms arranged in pentagons and hexagons, like a soccer ball, or graphite-like sheets rolled into tubes.

22.5   Natural graphite is still used in pencils, but modern uses are more varied. Graphite is one of the most studied materials for use as the negative electrode in batteries. It is also used in the construction of artificial heart valves.

22.6   Atoms can be trapped inside a buckminsterfullerene, so it can act as a special storage device. With a potassium atom inside, they show the properties of a superconductor at about 18 K. They also show catalytic activity. Nanotubes can be used to construct molecular-scale tweezers and possibly computer memory devices

22.7   Catenation is the ability of an atom to bond covalently to like atoms, as in ethylene, $H_2C=CH_2$.

22.8 $CO_2(g) + H_2O(l) \rightleftharpoons H_2CO_3(aq)$

$H_2CO_3(aq) \rightleftharpoons H^+(aq) + HCO_3^-(aq)$

$HCO_3^-(aq) \rightleftharpoons H^+(aq) + CO_3^{2-}(aq)$

22.9   Certain bacteria in the soil and in the roots of plants convert N_2 to ammonium and nitrate compounds. The plants then use these nitrogen compounds to make proteins and other complex nitrogen compounds. Animals eat the plants. Ultimately, the animals die, and bacteria in the decaying organic matter convert the nitrogen compounds back to N_2.

22.10   These oxides are nitrous oxide, N_2O; nitric oxide, NO; dinitrogen trioxide, N_2O_3; nitrogen dioxide, NO_2; dinitrogen tetroxide, N_2O_4; and dinitrogen pentoxide, N_2O_5. The oxidation numbers in each are +1, +2, +3, +4, +4, and +5, respectively.

22.11 Ammonia is burned in the presence of a platinum catalyst to form NO, which is then reacted with O_2 to form NO_2. The NO_2 is dissolved in water to form HNO_3 and NO. The NO is recycled back to the second step to react with more O_2 to form more NO_2.

22.12 $P_4(s) + 5O_2(g) \rightarrow P_4O_{10}(s)$

$P_4O_{10}(s) + 6H_2O(l) \rightarrow 4H_3PO_4(aq)$

22.13 The most important commercial means of producing oxygen is by distillation of liquid air. Air is filtered to remove dust particles, cooled to freeze out water and carbon dioxide, liquefied, and finally warmed until nitrogen and argon distill, leaving liquid oxygen behind.

22.14 Oxides are binary oxygen compounds in which oxygen is in the -2 oxidation state, whereas in peroxides, the oxidation number of oxygen is -1, and the anion is O_2^{2-}. In the superoxides, the oxidation number of oxygen is -1/2, and the anion is O_2^-. An example of each is H_2O (oxide), H_2O_2 (peroxide), and KO_2 (superoxide).

22.15 CrO_3 is an example of an acidic oxide; Cr_2O_3, MgO, and Fe_3O_4 are basic oxides.

22.16 First step: $S_8(s) + 8O_2(g) \rightarrow 8SO_2(g)$

Second step: $2SO_2(g) + O_2(g) \rightarrow 2SO_3(g)$

Third step: $SO_3(g) + H_2O(l) \rightarrow H_2SO_4(aq)$

22.17 a. $I_2(aq) + Cl^-(aq) \rightarrow NR$

b. $Cl_2(aq) + 2Br^-(aq) \rightarrow Br_2(aq) + 2Cl^-(aq)$

c. $Br_2(aq) + 2I^-(aq) \rightarrow I_2(s) + 2Br^-(aq)$

d. $Br_2(aq) + Cl^-(aq) \rightarrow NR$

22.18 Chlorine is used in preparing chlorinated hydrocarbons, as a bleaching agent, and as a disinfectant.

22.19 Sodium hypochlorite is prepared by reacting chlorine with NaOH:

$Cl_2(g) + 2NaOH(aq) \rightarrow NaClO(aq) + NaCl(aq) + H_2O(l)$

22.20 Bartlett found that PtF_6 reacted with molecular oxygen. Because the first ionization energy of xenon was slightly less than that of molecular oxygen, he reasoned that PtF_6 ought to react with xenon also.

■ Solutions to Practice Problems

Note on significant figures: If the final answer to a solution needs to be rounded off, it is given first with one nonsignificant figure, and the last significant figure is underlined. The final answer is then rounded to the correct number of significant figures. In multiple-step problems, intermediate answers are given with at least one nonsignificant figure; however, only the final answer has been rounded off.

22.25   For 2.5×10^4 kg (2.5×10^7 g) of hydrogen, you get

$$2.5 \times 10^7 \text{ g } H_2 \times \frac{1 \text{ mol } H_2}{2.016 \text{ g } H_2} \times \frac{-484 \text{ kJ}}{2 \text{ mol } H_2} = -2.\underline{9}98 \times 10^9 = -3.0 \times 10^9 \text{ kJ}$$

Thus 3.0×10^9 kJ of heat is evolved in the process.

22.27 a. The oxidation state of H in CaH_2 is -1.

b. The oxidation state of H in H_2O is +1.

c. The oxidation state of C in CH_4 is -4.

d. The oxidation state of S in H_2SO_4 is +6.

22.29 a. Carbon has four valence electrons. These are directed tetrahedrally, and the orbitals should be sp^3 hybrid orbitals. Each C–H bond is formed by the overlap of a $1s$ orbital of a hydrogen atom with one of the singly occupied sp^3 hybrid orbitals of the carbon atom.

b. Carbon has four valence electrons. The single C–C bond is a sigma bond formed by the overlap of the sp^2 orbital of the middle carbon and the sp^3 orbital of the CH_3 carbon. The other three orbitals of the CH_3 carbon are used to form the C–H bonds by overlapping with the orbital of the hydrogen atom. The C=C double bond is a sigma bond formed by the overlap of the sp^2 hybrid orbitals of those carbon atoms, and a pi bond formed by the overlap of the unhybridized p orbitals. The C–H bonds on the middle carbon and the CH_2 carbon are formed by the overlap of the sp^2 orbital of carbon and the s orbital of hydrogen.

22.31 a. $CO_2(g) + Ba(OH)_2(aq) \rightarrow BaCO_3(s) + H_2O(l)$

b. $MgCO_3(s) + 2HBr(aq) \rightarrow CO_2(g) + H_2O(l) + MgBr_2(aq)$

22.33 The density of diamond is 3.5155 g/cm^3, and the density of graphite is 2.2670 g/cm^3 (*CRC Handbook of Chemistry and Physics*, 71st ed). Therefore diamond is more dense than graphite. According to Le Châtelier's principle, as the pressure increases, the volume decreases, and higher densities are favored. Therefore, at higher pressures, diamond is the more stable allotrope.

22.35 Assume 1.00 mg of diamond is 1.00 mg (1.00×10^{-3} g) of carbon. Convert this to moles of methane

$$1.00 \times 10^{-3} \text{ g C} \times \frac{1 \text{ mol C}}{12.01 \text{ g C}} \times \frac{1 \text{ mol CH}_4}{1 \text{ mol C}} = 8.3\underline{2}63 \times 10^{-5} \text{ mol CH}_4$$

At STP, one mole of a gas occupies 22.4 L, so the volume of methane is

$$V = 8.3263 \times 10^{-5} \text{ mol} \times 22.4 \text{ L/mol} = 1.8\underline{6}51 \times 10^{-3} \text{ L} = 1.87 \text{ mL}$$

22.37 The equation is $Mg_3N_2(s) + 6H_2O(l) \rightarrow 3Mg(OH)_2(aq) + 2NH_3(g)$. Using a 1:2 ratio of Mg_3N_2 to NH_3, calculate the mass of NH_3 formed as follows:

$$7.50 \text{ g Mg}_3N_2 \times \frac{1 \text{ mol Mg}_3N_2}{100.915 \text{ g Mg}_3N_2} \times \frac{2 \text{ mol NH}_3}{1 \text{ mol Mg}_3N_2} \times \frac{17.03 \text{ g NH}_3}{1 \text{ mol NH}_3}$$

$$= 2.5\underline{3}1 = 2.53 \text{ g}$$

22.39 The reduction of NO_3^- ion to NH_4^+ ion is an eight-electron reduction, and the oxidation of zinc to zinc(II) ion is a two-electron oxidation. Balancing the equation involves multiplying the zinc half-reaction by four to achieve an eight-electron oxidation. The final equation is

$$4Zn(s) + NO_3^-(aq) + 10H^+(aq) \rightarrow 4Zn^{2+}(aq) + NH_4^+(aq) + 3H_2O(l)$$

22.41 PBr_4^+ has four pairs of electrons around the P atom, arranged in a tetrahedral fashion. The hybridization of P is sp^3. Each bond is formed by the overlap of a $4p$ orbital from Br with an sp^3 hybrid orbital from P.

22.43 a. $4Li(s) + O_2(g) \rightarrow 2Li_2O(s)$

b. Organic materials burn in excess O_2 to give CO_2 and H_2O. The nitrogen becomes N_2:

$$4CH_3NH_2(g) + 9O_2(g) \rightarrow 4CO_2(g) + 2N_2(g) + 10H_2O(g)$$

22.45 a. $x_S + 6x_F = 0$

The oxidation number of F in compounds is always -1.

$$x_S = -6x_F = -6(-1) = +6$$

b. $x_S + 3x_O = 0$

The oxidation number of O in most compounds is -2.

$$x_S = -3x_O = -3(-2) = +6$$

c. $x_S + 2x_H = 0$

The oxidation number of H in most compounds is +1.

$$x_S = -2x_H = -2(+1) = -2$$

d. $x_{Ca} + x_S + 3x_O = 0$

The oxidation number of Ca in compounds is +2; the oxidation number of O in most compounds is -2.

$$x_S = -x_{Ca} - 3x_O = -(+2) - 3(-2) = +4$$

22.47 $Ba(ClO_3)_2(aq) + H_2SO_4(aq) \rightarrow 2HClO_3(aq) + BaSO_4(s)$

22.49 a. The electron-dot formula of Cl_2O is

The VSEPR model predicts a bent (angular) molecular geometry. You can describe the four electron pairs on O using sp^3 hybrid orbitals. The diagramming for the bond formation follows:

(continued)

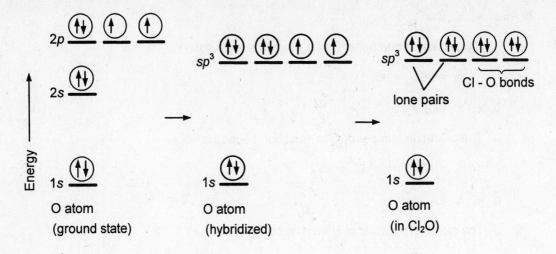

O atom
(ground state)

O atom
(hybridized)

O atom
(in Cl_2O)

b. An electron-dot formula of BrO_3^- is

$$
\left[\begin{array}{c} :\!\ddot{O}\!: \\ :\!Br\!:\!\ddot{O}\!: \\ :\!\ddot{O}\!: \end{array} \right]^-
$$

The VSEPR model predicts a trigonal pyramidal geometry. You can describe the four electron pairs on Br using sp^3 hybrid orbitals. The diagramming for the bond formation follows:

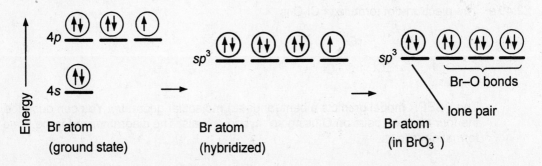

Br atom
(ground state)

Br atom
(hybridized)

Br atom
(in BrO_3^-)

(Note: The additional electron accounts for the -1 charge of the ion; the bonds to O atoms are coordinate covalent.)

(continued)

c. The electron-dot formula of BrF_3 is

The five electron pairs on Br have a trigonal bipyramidal arrangement. Putting the lone pairs in equatorial positions to reduce repulsions gives a T-shaped molecular geometry for BrF_3. You can describe the five electron pairs on Br in terms of sp^3d hybrid orbitals. The diagramming for the bond formation follows:

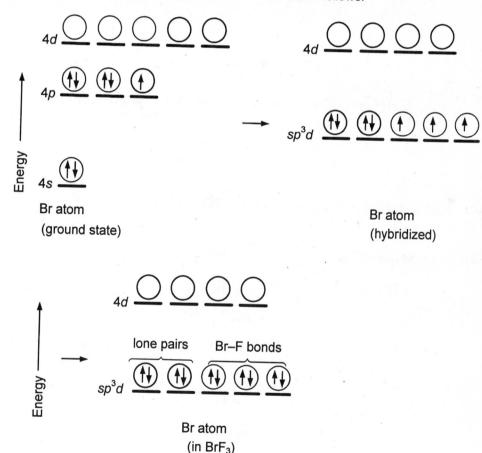

22.51 a. $Br_2(aq) + 2NaOH(aq) \rightarrow BrO^-(aq) + Br^-(aq) + H_2O(l) + 2Na^+(aq)$

b. Assume by analogy with H_2SO_4 and NaCl that the usual heating causes a loss of one H^+ per molecule of acid. Stronger heating would result in a loss of additional H^+ from $H_2PO_4^-$.

$$NaBr(s) + H_3PO_4(aq) \xrightarrow{\Delta} HBr(g) + NaH_2PO_4(aq)$$

22.53 The total number of valence electrons is $8 + (4 \times 7) = 36$. These are distributed to give the following Lewis formula:

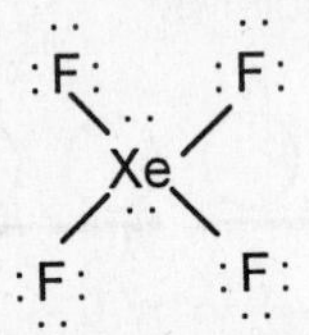

The six electron pairs on Xe would have an octahedral arrangement, suggesting sp^3d^2 hybridization. The lone pairs on Xe would be directed above and below the molecule, which has a square planar geometry.

■ Solutions to General Problems

22.55 The equations are

$$2HCl + Mg(OH)_2 \rightarrow MgCl_2 + 2H_2O$$

$$HCl + NaOH \rightarrow NaCl + H_2O.$$

Using a 1:1 ratio of HCl to NaOH, calculate the mol of HCl reacting with NaOH:

$$\left[\frac{0.4987 \text{ mol HCl}}{L} \times 0.05000 \text{ L} \right] - \left[\frac{0.2456 \text{ mol NaOH}}{L} \times 0.03942 \text{ L} \right]$$

$$= 0.015253 \text{ mol HCl}$$

(continued)

The mass of $Mg(OH)_2$ is

$$0.015253 \text{ mol HCl} \times \frac{1 \text{ mol } Mg(OH)_2}{2 \text{ mol HCl}} \times \frac{58.33 \text{ g } Mg(OH)_2}{1 \text{ mol } Mg(OH)_2}$$

$$= 0.44486 \text{ g } Mg(OH)_2$$

The mass percentage is

$$\frac{0.44486 \text{ g}}{5.436 \text{ g}} \times 100\% = 8.1837 = 8.184 \text{ percent}$$

22.57 From the ideal gas law, the moles of CO_2 are

$$n = \frac{PV}{RT} = \frac{(745/760 \text{ atm})(0.03456 \text{ L})}{(0.0821 \text{ L} \cdot \text{atm/K} \cdot \text{mol})(294 \text{ K})} = 0.0014035 \text{ mol } CO_2$$

The equation is $CaCO_3(s) + 2HCl(aq) \rightarrow CO_2(g) + CaCl_2(aq) + H_2O(l)$. Using a 1:1 mole ratio of CO_2 to $CaCO_3$, the mass of $CaCO_3$ is

$$0.0014035 \text{ mol } CO_2 \times \frac{1 \text{ mol } CaCO_3}{1 \text{ mol } CO_2} \times \frac{100.1 \text{ g } CaCO_3}{1 \text{ mol } CaCO_3} = 0.14049 \text{ g } CaCO_3$$

The mass percentage of $CaCO_3$ is

$$\frac{0.14049 \text{ g}}{0.1662 \text{ g}} \times 100\% = 84.53 = 84.5 \text{ percent } CaCO_3$$

22.59 Using the ideal gas law, the amount of CO_2 is

$$n = \frac{PV}{RT} = \frac{(30.0/760 \text{ atm})(1.00 \text{ L})}{(0.0821 \text{ L} \cdot \text{atm/K} \cdot \text{mol})(298 \text{ K})} = 1.614 \times 10^{-3} \text{ mol } CO_2$$

The balanced equation is $2LiOH(s) + CO_2(g) \rightarrow Li_2CO_3(s) + H_2O(g)$, so use the mole ratio one mol CO_2 to two mol LiOH. Therefore the mass of LiOH is

$$1.614 \times 10^{-3} \text{ mol } CO_2 \times \frac{2 \text{ mol LiOH}}{1 \text{ mol } CO_2} \times \frac{23.95 \text{ g LiOH}}{1 \text{ mol LiOH}}$$

$$= 7.731 \times 10^{-2} = 7.73 \times 10^{-2} \text{ g LiOH}$$

22.61   Assume a sample of 100.0 g fertilizer. This contains 17.1 g P. Convert this to the mass of $Ca(H_2PO_4)_2 \cdot H_2O$ in the 100.0 g of fertilizer.

$$17.1 \text{ g P} \times \frac{1 \text{ mol P}}{30.97 \text{ g P}} \times \frac{1 \text{ mol Ca(H}_2\text{PO}_4)_2 \cdot H_2O}{2 \text{ mol P}}$$

$$\times \frac{252.1 \text{ g Ca(H}_2\text{PO}_4)_2 \cdot H_2O}{1 \text{ mol Ca(H}_2\text{PO}_4)_2 \cdot H_2O} = 69.\underline{5}89 \text{ g Ca(H}_2\text{PO}_4)_2$$

The mass percent $Ca(H_2PO_4)_2 \cdot H_2O$ in the fertilizer is

$$\text{Mass percent Ca(H}_2\text{PO}_4)_2 \cdot H_2O = \frac{69.589 \text{ g Ca(H}_2\text{PO}_4)_2 \cdot H_2O}{100 \text{ g fertilizer}} \times 100\%$$

$$= 69.\underline{5}89 = 69.6 \text{ percent}$$

22.63 $2NaCl + 2H_2O \rightarrow 2NaOH + H_2 + Cl_2$ (electrolysis)

$2NaOH + Cl_2 \rightarrow NaClO + NaCl + H_2O$ (spontaneous)

One mol of NaClO is produced for every two mol of NaCl. Thus the reaction requires two electrons per mol of NaClO produced.

Convert the 1.00×10^3 L of NaOCl to time:

$$1.00 \times 10^3 \text{ L} \times \frac{1000 \text{ mL}}{1 \text{ L}} \times \frac{1.00 \text{ g soln}}{1 \text{ mL}} \times \frac{5.25 \text{ g NaOCl}}{100 \text{ g soln}} \times \frac{1 \text{ mol NaOCl}}{74.44 \text{ g NaOCl}}$$

$$\times \frac{2 \text{ mol e}^-}{1 \text{ mol NaOCl}} \times \frac{9.65 \times 10^4 \text{ C}}{1 \text{ mol e}^-} \times \frac{1 \text{ s}}{3.00 \times 10^3 \text{ C}} \times \frac{1 \text{ h}}{3600 \text{ s}}$$

$$= 12.\underline{6}0 = 12.6 \text{ h}$$

22.65   Graphite is a covalent network solid with a layered structure. The layers are attracted to one another by London forces. Within each layer, each carbon is covalently bonded to three other carbon atoms, in hexagons, with sp^2 hybridization and delocalized π bonds. Fullerenes are molecules consisting of a closed cage of carbon atoms arranged in pentagons and hexagons. Other fullerenes consist of graphite-like sheets of carbon atoms rolled into tubes.

They are similar in that they comprise only carbon atoms with sp^2 hybridization, each attached to three other carbon atoms and also with delocalized π bonds. They are different in that fullerenes can have pentagons and hexagons but graphite has only hexagons. Also, fullerenes are distinct molecules while graphite is a covalent network solid.

23. ORGANIC CHEMISTRY

■ Answers to Review Questions

23.1   The formula of an alkane with 20 carbon atoms is $C_{20}H_{42}$.

23.2 $CH_3-CH_2-CH_2-CH_2-CH_3$

$CH_3-\underset{\underset{\displaystyle CH_3}{|}}{CH}-CH_2-CH_3$

$CH_3-\underset{\underset{\displaystyle CH_3}{|}}{\overset{\overset{\displaystyle CH_3}{|}}{CH}}-CH_3$

23.3   The two isomers of 3-pentene are the *cis* and *trans* geometric isomers:

cis-3-pentene   *trans*-3-pentene

In the *cis*-3-pentene, the two groups are on the same side of the double bond; in the *trans* isomer, they are on opposite sides.

23.4 The structural formulas for the isomers of ethyl-methylbenzene are

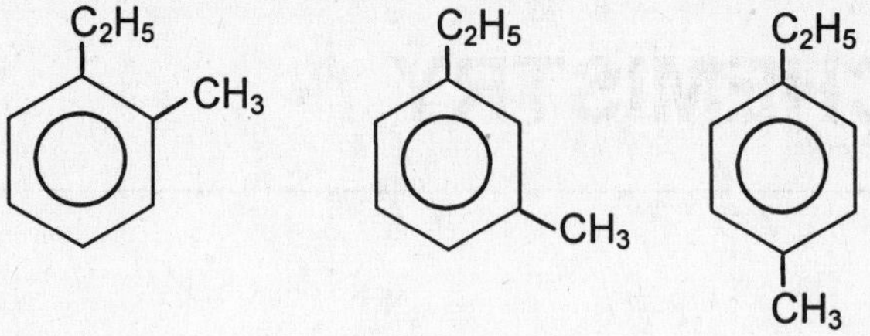

23.5

CH_3CH_2Br	CH_2BrCH_2Br	$CHBr_2CHBr_2$	CBr_3CBr_3
CH_3CHBr_2	$CH_2BrCHBr_2$	$CHBr_2CBr_3$	
CH_3CBr_3	CH_2BrCBr_3		

23.6 A substitution reaction is a reaction in which part of the reagent molecule is substituted for a hydrogen atom on a hydrocarbon or hydrocarbon group. For example,

$$CH_4 + Cl_2 \rightarrow CH_3Cl + HCl$$

An addition reaction is a reaction in which parts of the reagent are added to each carbon atom of a carbon-carbon multiple bond, which then becomes a C—C single bond. For example,

$$CH_2 = CH_2 + Br_2 \rightarrow CH_2Br–CH_2Br$$

23.7 The major product of HBr plus acetylene should be $Br_2HC–CH_3$ because Markownikoff's rule predicts this.

23.8 The structures are

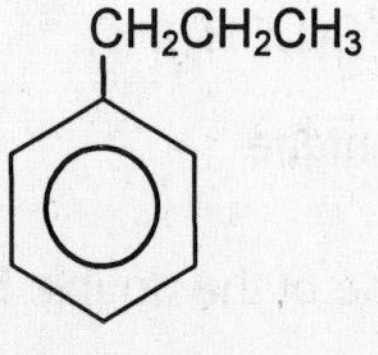

propylbenzene

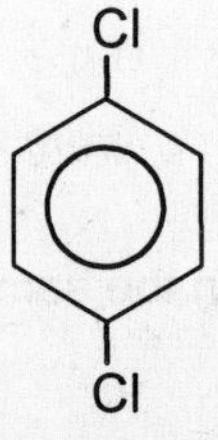

paradichlorobenzene

23.9 A functional group is a reactive portion of a molecule that undergoes predictable reactions no matter what the rest of the molecule is like. An example is a C=C bond, which always reacts with bromine or other addition reagents to add part of each reagent to each carbon atom.

23.10 An aldehyde is different from a ketone, carboxylic acid, and ester in that a hydrogen atom is always attached to the carbonyl group in addition to a hydrocarbon group.

23.11 a. CO is a carbonyl group (ketone).

b. CH_3O—C is an ether group.

c. C=C is a double bond.

d. COOH is a carboxylic acid group.

e. CHO is an aldehyde (carbonyl).

f. CH_2OH, or -OH, is a hydroxyl group (primary alcohol).

23.12 An addition polymer is a polymer formed by linking together many molecules by addition reactions. The monomers have multiple bonds that will undergo addition reactions. An example is the formation of polypropylene from propene.

A condensation polymer is formed by linking together many molecules by condensation reactions. An example is the formation of nylon.

■ Solutions to Practice Problems

23.19 The condensed structural formula is $CH_3CH_2CH_2CH_3$.

23.21 a.

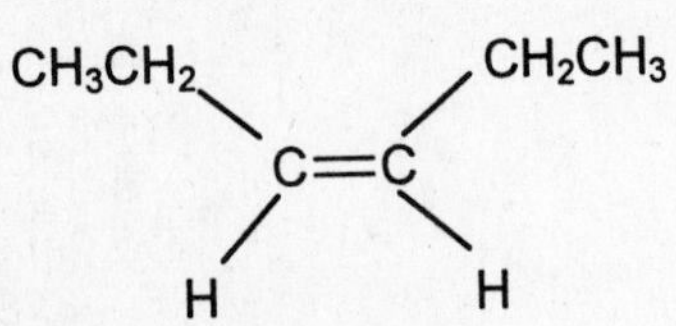

cis-3-hexene

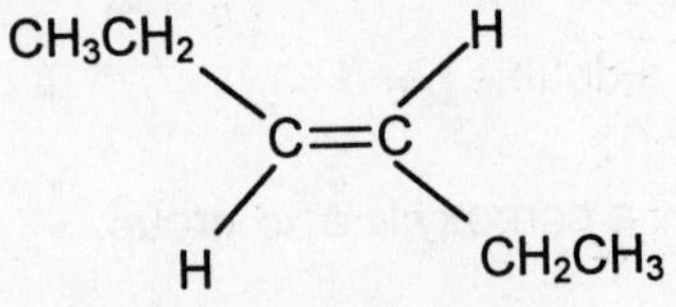

trans-3-hexene

b.

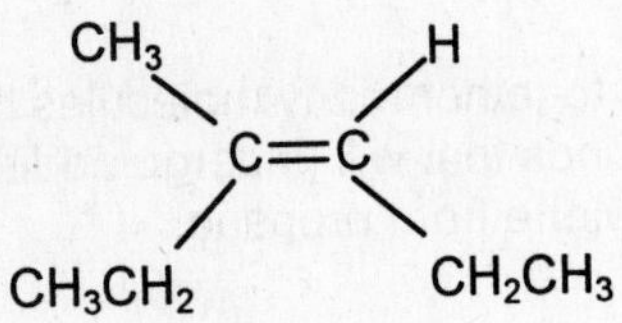

cis-3-methyl-3-hexene

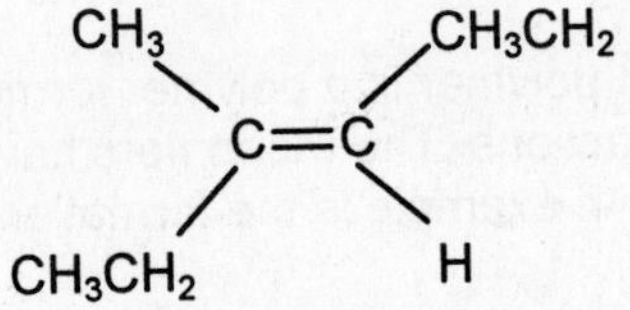

trans-3-methyl-3-hexene

23.23 a. $C_2H_4 + 3O_2 \rightarrow 2CO_2 + 2H_2O$

b. $CH_2{=}CH_2 + MnO_4^- + H_2O \rightarrow \overset{\text{OH OH}}{CH_2CH_2} + MnO_2$

Oxidation: $CH_2{=}CH_2 \rightarrow \overset{\text{OH OH}}{CH_2CH_2}$

Balance O: $2OH^- + CH_2{=}CH_2 \rightarrow \overset{\text{OH OH}}{CH_2CH_2}$

Balance e⁻ $2OH^- + CH_2{=}CH_2 \rightarrow \overset{\text{OH OH}}{CH_2CH_2} + 2e^-$

Reduction: $MnO_4^- \rightarrow MnO_2$

Balance O: $MnO_4^- + 2H_2O \rightarrow MnO_2 + 4OH^-$

Balance e⁻: $MnO_4^- + 2H_2O + 3e^- \rightarrow MnO_2 + 4OH^-$

(continued)

Add half-reactions:

$$3\left(2OH^- + CH_2{=}CH_2 \rightarrow \overset{\displaystyle OH \;\; OH}{\underset{\displaystyle CH_2CH_2}{|\;\;\;\;|}} + 2e^-\right)$$

$$2\left(MnO_4^- + 2H_2O + 3e^- \rightarrow MnO_2 + 4OH^-\right)$$

$$\cancel{6}OH^- + 3CH_2{=}CH_2 + 2MnO_4^- + 4H_2O \rightarrow 3\,\overset{\displaystyle OH\;\;OH}{\underset{\displaystyle CH_2CH_2}{|\;\;\;\;|}} + 2MnO_2 + \overset{2}{\cancel{8}}OH^-$$

$$3CH_2{=}CH_2 + 2MnO_4^- + 4H_2O \rightarrow 3\,\overset{\displaystyle OH\;\;OH}{\underset{\displaystyle CH_2CH_2}{|\;\;\;\;|}} + 2MnO_2 + 2OH^-$$

c. $CH_2{=}CH_2 + Cl_2 \rightarrow \overset{\displaystyle Cl\;\;Cl}{\underset{\displaystyle CH_2CH_2}{|\;\;\;\;|}}$

d.

23.25 According to Markownikoff's rule, the major product is the one obtained when the H atom adds to the carbon atom of the double bond that already has more hydrogen atoms attached to it. Therefore 2-chloro-2-methylpropane is the major product.

23.27 a.

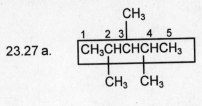

            longest chain   2,3,4-trimethylpentane

b.

$$\begin{array}{c} CH_3 CH_3 \\ \underset{1}{} \underset{2}{} \underset{3}{} \underset{4}{} \underset{5}{} \underset{6}{} \underset{7}{} \\ \boxed{CH_3CCH_2CH_2CH_2CCH_3} \\ CH_3 CH_3 \end{array}$$

            longest chain   2,2,6,6-tetramethylheptane

c.

$$\begin{array}{c} \underset{4}{} \underset{5}{} \underset{6}{} \underset{7}{} \underset{8}{} \\ CH_3CH_2\boxed{CHCH_2CH_2CH_2CH_3} \\ \boxed{CH_2CH_2CH_3} \\ \underset{3}{} \underset{2}{} \underset{1}{} \end{array}$$

            longest chain   4-ethyloctane

23.29 a.

$$\begin{array}{c} CH_3 \\ | \\ CH_3CHCHCH_2CH_2CH_2CH_3 \\ | \\ CH_3 \end{array}$$

b.

$$\begin{array}{c} CH_2CH_3 \\ | \\ CH_3CH_2CHCH_2CH_2CH_3 \end{array}$$

c.

$$\begin{array}{c} CH_3 \\ | \\ CH_3CHCH_2CHCH_2CH_2CH_3 \\ CH_3CHCH_3 \end{array}$$

23.31 a. 2-pentene

     b. 2,5-dimethyl-2-hexene

23.33 a.

$$\begin{array}{c} CH_3CH{=}CCH_2CH_3 \\ | \\ CH_2CH_3 \end{array}$$

b.

$$\begin{array}{c} CH_3C{=}CHCHCH_2CH_3 \\ | | \\ CH_3 CH_2CH_3 \end{array}$$

23.35   The two isomers are *cis*-2-pentene and *trans*-2-pentene.

23.37 a. 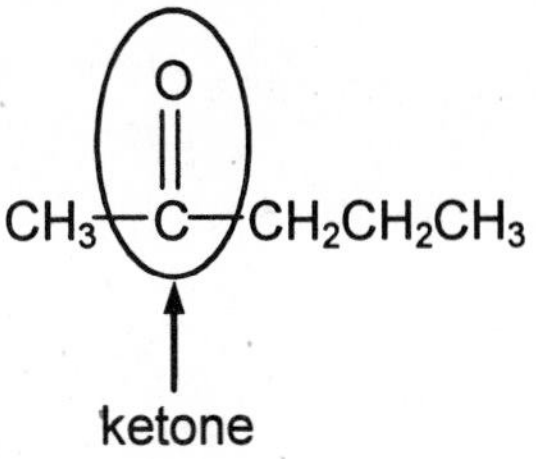 CH_3—C—$CH_2CH_2CH_3$
(O double bond, circled) ketone

b. 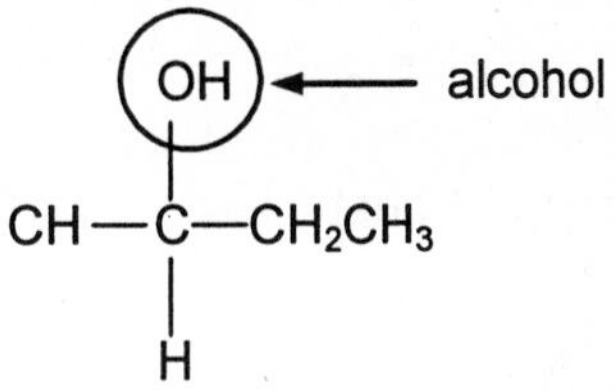   CH—C—CH_2CH_3
(OH ← alcohol, circled; H below)

c.   HO—C—CH_2CH_3
(O double bond, circled) carboxylic acid

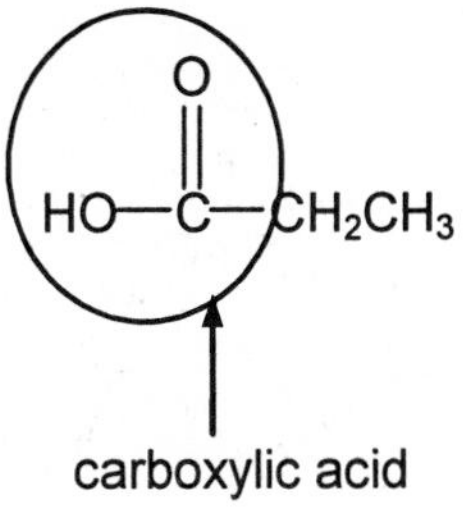

d. 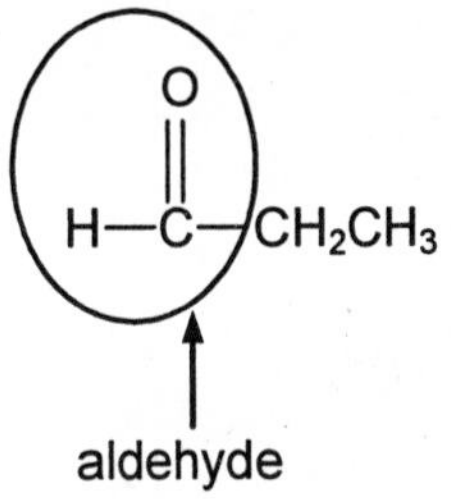   H—C—CH_2CH_3
(O double bond, circled) aldehyde

23.39 a.   1-pentanol

b.   2-pentanol

c.   2-propyl-1-pentanol

23.41 a.   secondary alcohol

b.   secondary alcohol

c.   primary alcohol

d.   primary alcohol

23.43 a.   butanone

b.   butanal

c.   4,4-dimethylpentanal

d.   3-methyl-2-pentanone

23.45 a.   secondary amine

b.   secondary amine

■ Solutions to General Problems

23.47 a. 3-methylbutanoic acid   **b.** *trans*-5-methyl-2-hexene

c. 2,5-dimethyl-4-heptanone   **d.** 4-methyl-2-pentyne

23.49 a.

$$CH_3CH_2\overset{\overset{\displaystyle O}{\|}}{C}-O-CH_2CH_2CH_3$$

b.

$$CH_3-\overset{\overset{\displaystyle CH_3}{|}}{\underset{\underset{\displaystyle CH_3}{|}}{CH}}-NH_2$$

c.

$$CH_3CH_2CH_2\overset{\overset{\displaystyle CH_3}{|}}{\underset{\underset{\displaystyle CH_3}{|}}{C}}-COOH$$

d.

$$\underset{CH_3CH_2}{\overset{H}{\diagdown}}C=C\underset{H}{\overset{CH_2CH_3}{\diagup}}$$

23.51 a. Addition of dichromate ion in acidic solution to propionaldehyde will cause the reagent to change from orange to green as the aldehyde is oxidized. Under similar conditions, acetone (a ketone) would not react.

b. Addition of a solution of Br_2 in CCl_4 to $CH_2=CH-C\equiv C-CH=CH_2$ would cause the bromine color to disappear as the Br_2 was added to the multiple bonds. Addition of benzene to Br_2 in CCl_4 results in no reaction. Aromatic rings are not susceptible to attack by Br_2 in the absence of a catalyst.

23.53 a. ethylene, $CH_2=CH_2$

b. There must be an aromatic ring in the compound or the double bonds would react with Br_2. The compound is toluene, and the correct formula is

c. methylamine, CH_3NH_2

d. methanol, CH_3OH

23.55 $nCF_2=CF_2 \rightarrow -CF_2-CF_2-CF_2-CF_2-CF_2-CF_2-$

23.57 The two monomer units for this polymer are

$$HOCH_2CH_2OH \quad and \quad HO\overset{\overset{O}{\|}}{C}CH_2CH_2\overset{\overset{O}{\|}}{C}OH$$

23.59 Assume 100.0 g of the unknown. This contains 85.6 g C and 14.4 g H. Convert these amounts to moles.

$$85.6 \text{ g} \times \frac{1 \text{ mol C}}{12.01 \text{ g C}} = 7.1\underline{2}7 \text{ mol C}$$

$$14.4 \text{ g H} \times \frac{1 \text{ mol H}}{1.008 \text{ g H}} = 14.\underline{2}857 \text{ mol H}$$

The molar ratio of H to C is 14.2857:7.12, or 2.00:1. The empirical formula is therefore CH_2. This formula unit has a mass of $[12.01 + 2 \times (1.008)]$ amu = 14.026 amu.

$$\frac{56.1 \text{ amu}}{1 \text{ molecule}} \times \frac{1 \text{ formula unit}}{14.026 \text{ amu}} = \frac{4.00 \text{ formula units}}{1 \text{ molecule}}$$

The molecular formula is $(CH_2)_4$, or C_4H_8. The formula (C_nH_{2n}) indicates the compound is either an alkene or a cycloalkane. Because it reacts with water and H_2SO_4, it must be an alkene. The product of the addition of H_2O to a double bond is an alcohol. Because the alcohol produced can be oxidized to a ketone, it must be a secondary alcohol. The only secondary alcohol with four carbon atoms is 2-butanol.

$$CH_3CH_2\underset{\underset{OH}{|}}{C}HCH_3$$

The original hydrocarbon from which it was produced is either 1-butene or 2-butene.

$$CH_3CH=CHCH_3 + H_2O \xrightarrow{H_2SO_4} CH_3CH_2\underset{\underset{OH}{|}}{C}HCH_3$$

or

$$CH_3CH_2CH=CH_2 + H_2O \xrightarrow{H_2SO_4} CH_3CH_2\underset{\underset{OH}{|}}{C}HCH_3$$

APPENDIX A. MATHEMATICAL SKILLS

■ Solutions to Exercises

1. a. Either leave it as 4.38 or write it as 4.38×10^0.

    b. Shift the decimal point left and count the number of positions shifted (3). The answer is 4.380×10^3 assuming the terminal zero is significant.

    c. Shift the decimal point right and count the number of positions shifted (4). The answer is 4.83×10^{-4}.

2. a. Shift the decimal point right three places. The answer is 7025.

    b. Shift the decimal point left four places. The answer is 0.000897.

3. Express 2.8×10^{-6} as 0.028×10^{-4}. Then the sum can be written

    $$(3.142 \times 10^{-4}) + (0.028 \times 10^{-4}) \text{ or } (3.142 + 0.028) \times 10^{-4} = 3.170 \times 10^{-4}.$$

4. a. $(5.4 \times 10^{-7}) \times (1.8 \times 10^8) = (5.4 \times 1.8) \times 10^{-7} \times 10^8 = 9.72 \times 10^1$. This rounds to 9.7×10^1.

    b. $\dfrac{5.4 \times 10^{-7}}{6.0 \times 10^{-5}} = \dfrac{5.4}{6.0} \times 10^{-7} \times 10^5 = 0.90 \times 10^{-2} = 9.0 \times 10^{-3}$

5. a. $(3.56 \times 10^3)^4 = (3.56)^4 \times (10^3)^4 = 161 \times 10^{12} = 1.61 \times 10^{14}$

    b. $\sqrt[3]{4.81 \times 10^2} = \sqrt[3]{0.481 \times 10^3} = \sqrt[3]{0.481} \times \sqrt[3]{10^3} = 0.784 \times 10^1 = 7.84 \times 10^0$

6. a. log 0.00582 = -2.235

    b. log 689 = 2.838

7. a. antilog 5.728 = 5.35×10^5

    b. antilog (-5.728) = 1.87×10^{-6}

8. $$x = \frac{-0.850 \pm \sqrt{(0.850)^2 - 4(1.80)(-9.50)}}{2(1.80)} = \frac{-0.850 \pm 8.314}{3.60}$$

    The positive root is

    $$\frac{7.46}{3.60} = 2.07$$